Patrick Barry

The Fruit Garden

Patrick Barry

The Fruit Garden

ISBN/EAN: 9783337068806

Printed in Europe, USA, Canada, Australia, Japan

Cover: Foto ©berggeist007 / pixelio.de

More available books at **www.hansebooks.com**

THE

FRUIT GARDEN;

A TREATISE

INTENDED TO EXPLAIN AND ILLUSTRATE THE PHYSIOLOGY OF FRUIT
TREES, THE THEORY AND PRACTICE OF ALL OPERATIONS

CONNECTED WITH THE

PROPAGATION, TRANSPLANTING, PRUNING AND TRAINING OF
ORCHARD AND GARDEN TREES, AS STANDARDS, DWARFS,
PYRAMIDS, ESPALIERS, ETC.,

THE LAYING OUT AND ARRANGING DIFFERENT KINDS OF

ORCHARDS AND GARDENS,

THE SELECTION OF SUITABLE VARIETIES FOR DIFFERENT PURPOSES AND
LOCALITIES, GATHERING AND PRESERVING FRUITS, TREATMENT OF
DISEASES, DESTRUCTION OF INSECTS, DESCRIPTIONS AND
USES OF IMPLEMENTS, ETC.

ILLUSTRATED WITH UPWARDS OF 150 FIGURES,

REPRESENTING DIFFERENT PARTS OF TREES, ALL PRACTICAL OPERATIONS,
FORMS OF TREES, DESIGNS FOR PLANTATIONS, IMPLEMENTS, ETC.

BY P. BARRY,
OF THE MOUNT HOPE NURSERIES, ROCHESTER, NEW YORK.

NEW YORK:
C. M. SAXTON, BARKER & CO., 25 PARK ROW.
SAN FRANCISCO: H. H. BANCROFT & CO.
1860.

Entered according to Act of Congress, in the year 1861, by
CHARLES SCRIBNER,
In the Clerk's Office of the District Court of the United States for the Southern District of New York.

INTRODUCTION.

The subject of this treatise is one in which almost all classes of the community are more or less practically engaged and interested. Agriculture is pursued by one class, and commerce by another; the mechanic arts, fine arts, and learned professions by others; but fruit culture, to a greater or less extent, by *all*.

It is the desire of every man, whatever may be his pursuit or condition in life, whether he live in town or country, to enjoy fine fruits, to provide them for his family, and, if possible, to cultivate the trees in his own garden with his own hands. The agriculturist, whatever be the extent or condition of his grounds, considers an *orchard*, at least, indispensable. The merchant or professional man who has, by half a lifetime of drudgery in town, secured a fortune or a competency that enables him to retire to a country or suburban villa, looks forward to his fruit garden as one of the chief sources of those rural comforts and pleasures he so long and so earnestly labored and hoped for. The artizan who has laid up enough from his earnings to purchase a homestead, considers the planting of his fruit trees as one of the first and most important steps towards improvement. He anticipates the pleasure of tending them in his spare hours, of watching their growth and progress to maturity, and of gathering their ripe and delicious fruits, and placing them before his family and friends as the valued products of his own garden, and of his own skill and labor. Fortunately, in the United States, land is so easily obtained as to be within the reach of every industrious man ; and the climate and soil being so favorable to the production of fruit, Americans, if they be not already, must become truly " a nation of fruit growers."

Fruit culture, therefore, whether considered as a branch of profitable industry, or as exercising a most beneficial influence upon the health, habits, and tastes of the people, becomes a great national interest, and whatever may assist in making it better understood, and more interesting, and better adapted to the various wants, tastes, and circumstances of the community, cannot fail to subserve the public good.

Within a few years past it has received an unusual degree of attention. Plantations of all sorts, orchards, gardens, and nurseries, have increased in numbers and extent to a degree quite unprecedented; not in one section or locality, but from the extreme north to the southern limits of the fruit-growing region. Foreign supplies of trees have been required to meet the suddenly and greatly increased demand. Treatises and periodicals devoted to the subject have increased rapidly and circulated widely. Horticultural societies have been organized in all parts; while exhibitions, and national, state, and local conventions of fruit growers, have been held to discuss the merits of fruits and other kindred topics.

To those unacquainted with the previous condition of fruit culture in the interior of the country, this new planting spirit has appeared as a sort of speculative mania, and the idea has suggested itself to them that the country will soon be overstocked with fruits. This is a greatly mistaken apprehension. After all that has been done, let us look at the actual condition of fruit culture at the present time. In the best fruit-growing counties in the State of New York, the entire fruit plantations of more than three fourths of the agricultural population consist of very ordinary orchards of apples. Not a dish of fine pears, plums, cherries, apricots, grapes, or raspberries, has ever appeared on their tables, and not a step has yet been taken to produce them. People are but beginning to learn the uses of fruits, and to appreciate their importance.

At one time apples were grown chiefly for cider; now they are considered indispensable articles of food. The finer fruits, that were formerly considered as luxuries only for the tables of the wealthy, are beginning to take their place among the ordinary supplies of every man's table; and this taste must grow from year to year, with an increased supply. Those who consume a bushel of fruit this year, will require double or treble that quantity next. The rapid increase of population alone, creates a demand to an extent that few people are aware of. The city of Rochester has added 20,000 to her numbers in ten years. Let such an increase as this in all our cities, towns, and villages, be estimated, and see what an aggregate annual amount of new consumers it presents.

New markets are continually presenting themselves and demanding large supplies. New and more perfect modes of packing and shipping fruits, and of drying, preserving, and preparing them for various purposes to which they have not hitherto been appropriated, are beginning to enlist attention and inquiry.

Immense amounts of money are annually expended in importing grapes, wines, figs, nuts, prunes, raisins, currants, almonds, &c., many of which might be produced perfectly well on our own soil. Pears have actually been imported from France by the New York confectioners, this present season, (1851.) These are facts that should be well understood by proprietors of lands, and especially by those who have allowed themselves to imagine that fruit will soon be so plenty as not to be worth the growing

It is too soon by a century to apprehend an over supply of fruits in the United States, except of some very perishable sort, in a season of unusual abundance, in some particular locality where one branch of culture is mainly carried on.

It is because fruit culture has been almost entirely neglected until within a few years, that the present activity appears so extraordinary. A vast majority of the people were quite unaware of the treasures within their reach; and that in regard to soil and climate they possessed advantages for fruit growing superior to any other nation We had no popular works or periodicals to diffuse information or awaken interest on the subject. For fourteen or fifteen years Hovey's *Magazine of Horticulture* was the only journal exclusively devoted to gardening subjects, and it only found its way into the hands of the more advanced cultivators. We had some treatises on fruits, but none of them circulated sufficiently to effect much good. Previous to 1845, *Kenrick's American Orchardist*, and *Manning's Book of Fruits*, were the principal treatises that had any circulation worth naming. Coxe's work, Floy's, Prince's, and some others, were confined almost wholly to nursery-men, or persons already engaged and interested in fruit culture in the older parts of the country.

Mr. Downing's "Fruit and Fruit Trees of America," that appeared in 1845, was the first treatise of the kind that really obtained a wide and general circulation.

It made its appearance at a favorable moment, just as the planting spirit referred to was beginning to manifest itself, and when, more than at any previous period, such a work was needed. Mr. Downing enjoyed great advantages over any previous American writer. During the ten years that had elapsed since the publication of Kenrick's and Prince's treatises, a great fund of materials had been accumulating. Messrs. Manning, Kenrick, Prince, Wilder, and many others, had been industriously collecting fruits both at home and abroad. The Massachusetts Horticultural Society was actively engaged in its labors. The London Horticultural Society had made great advancement in its examination and trial of fruits, and had corrected a multitude of long standing errors in nomenclature.

Mr. Downing's work had the benefit of all this; and possessing the instructive feature of outline figures of fruits, and being written in a very agreeable and attractive style, it possessed the elements of popularity and usefulness in an eminent degree. Hence it became at once the text-book of every man who sought for pomological information, or felt interested in fruits or fruit trees; and to it is justly attributable much of the taste and spirit on the subject, and the increased attention to nomenclature, that so distinguishes the present time. Mr. Thomas's recent treatise, "The American Fruit Culturist," on the same plan as Mr. Downing's, is also a popular work, and will be the means of diffusing both taste and information. Mr Thomas is a close and accurate observer, and his descriptions are peculiarly concise, methodical and minute. "Cole's Fruit Book" is also a recent treatise, and on account of its cheapness and the vast accumulation of facts and information it contains is highly popular and useful. Besides these, periodicals devoted more or less to the subject, have increased in number and greatly extended their circulation, so that information is now accessible to all who desire it.

The light which has been shed upon fruit-growing by these works, and the taste they have created, have not only improved old systems of cultivation, but introduced *new* ones. Until within a few years nothing was said or known among the great body of cultivators, or even nursery-men, of dwarfing trees, of the uses of certain stocks, or of modes of propagation and pruning by which trees are made to bear early, and are adapted to different circumstances. The entire routine of the propagation and management of trees was conducted generally in the simplest and rudest manner Whether for the garden or the orchard they were propagated in the same manner, on the same stocks, and in the same form taken from the nursery, planted out and left there to assume such forms as nature or accident might impose, and produce fruit at such a time as natural circumstances would admit.

The art of planting fifty trees on a quarter of an acre of ground, and bringing them into a fruitful state in four or five years at most, was entirely unknown. Small gardens were encumbered with tall, unshapely, and unfruitful trees, that afforded no pleasure to the cultivator; and thousands of persons, who are now the most enthusiastic cultivators, were entirely discouraged from the attempt

Fruit gardening, properly speaking, may be said only to have commenced. It is no longer a matter of mere utility, but of taste also; and, therefore, *adaptation, variety*, and *beauty*, are sought for in garden trees and modes of culture and management. Nothing so distinguishes the taste of modern planting as the partiality for dwarf trees, and the desire to obtain information in regard to their propagation and treatment.

This has not been anticipated by any of our authors. The standard or orchard system alone is fully treated of, as being the only one practised; and this requires so little skill in the art of culture, that only the simplest instructions have been given. The very elements of the science have been unexplained and unstudied, and cultivators in the main find themselves both destitute of knowledge in regard to the management of trees in the more refined and artificial forms, and the sources from which to obtain it. But a very small proportion of those engaged or engaging in tree culture have studied the physiology of trees in any degree. Very few have the slightest knowledge of the modes of growth and bearing of the different species of fruits, or even of the difference between wood or leaf buds, and fruit buds. Very few understand the functions of the different parts of trees, and the relation in which they stand one to another; the principles that govern and regulate the growth and maturity, the formation of wood and the production of fruit. *Practice* is no better understood than principle. Persons engaged largely in tree growing will frequently ask the most absurd questions on the subject of propagation, of stocks, of pruning, &c., matters that should be understood by every man who has a single tree to manage, but especially indispensable to those who wish to succeed in conducting garden trees under certain modified forms, more or less opposed to the natural. The preparation of ground, laying out small gardens, the selection of suitable trees, and a multitude of minor but nevertheless important matters, are very imperfectly understood. Neither our state nor national governments have ever manifested a disposition to favor the rural arts with anything like a liberal patronizing policy. Advanced, wealthy and powerful as we are, not a single step has been taken, in earnest, to establish model farms or model gardens, in which experiments might be made and examples given that would enlighten cultivators, and elevate and honor their profession. Whatever advance has been made is due wholly to individual taste, energy, and enterprise; and to these alone are we permitted to look for future progress.

Having for many years devoted much attention to this particular branch of culture, and feeling deeply interested in its success, and having, by a business intercourse with cultivators in all parts of the country, an ample opportunity of understanding the nature and extent of the information desired, I have prepared the following pages to supply it at least in part.

I am well convinced that the work is neither perfect nor complete. It has been prepared, during a few weeks of the winter, in the midst of other engagements that rendered it impossible to bestow upon it the necessary care and labor. My original intention was to give a few brief directions for the management of garden trees, but it was suggested by friends that it would prove more generally useful by adding a sketch of the entire routine of operations, from the propagation in the nursery to the management in the orchard and garden. This has involved much more labor than it was intended to bestow on it, or than I could really spare from business. It has, therefore, been performed hastily, and, of course, in many respects imperfectly, but yet it is hoped it contains such an exposition of principles and practices as cannot fail to diffuse amongst the inexperienced much needed information. All doubtful theories, and whatever had not a direct practical bearing on the subjects treated, have been excluded, both for the sake of brevity, and to avoid anything calculated to mislead. The principles and practices set forth are not new, visionary, or doubtful, but such as are taught and practised by the most accomplished cultivators of the day, and have been successfully carried out in the daily operations of our own establishment.

- In the pruning and management of garden trees, the French arboriculturists surpass all others. Their trees are models that have no equals, and that all the world admire. The English, notwithstanding their great gardening skill, and their refined and elegant modes of culture, are far behind the French in the management of fruit trees. French systems of pruning and training are at this moment advocated and held up as models by such men as Mr. Robert Thompson, head of the fruit department in the London Horticultural Society's Garden ; by Mr. Rivers, well known on this side of the Atlantic as one of the most energetic and accomplished nurserymen in Great Britain; and by many others whose skill and judgment command attention. Their introduction to English gardens is going on rapidly, and bids fair to revolutionize their whole practice of fruit tree culture.

D'Albret's great work on pruning is conceded to be the best extant on that subject. He was the pupil and successor of M. Thouin, the world-renowned vegetable physiologist and founder of the great national gardens at Paris. His practice is founded upon the true principles of vegetable physiology, and strengthened by long years of the most minute and successful experiment.

M. Dubrieul, late conductor of the fruit department in the Garden of Rouen, has also published an excellent treatise on arboriculture ; and there are many other French works on the subject, all showing how thoroughly the science is there understood, and how minutely and skilfully its principles are dealt with. These, as well as the best-managed gardens and the most perfect and beautiful trees in France and Belgium, have been carefully studied.

The knowledge thus acquired, added to the experience of many years' actual and extensive practice, constitutes the basis of the course recommended.

The same minute detail that characterizes European works has not been attempted, yet much detail is absolutely necessary in order to prevent misapprehension on the part of those wholly inexperienced.

Writers are apt to treat simple matters too much in the general, presuming them to be well understood. Detail is always tedious to those familiar with the subject, but nothing less can be satisfactory to the student.

For the sake of convenient reference, the different branches of the subject have been separated into four parts. The *first* treats of general principles, a knowledge of the structure, character, and functions of the different parts of trees, modes of growth, bearing, &c., &c.; soils, manures, modes of propagation, &c. This must be the ground-work of the study of tree culture The *second* treats of the nursery. The *third* of plantations, orchards of different kinds, gardens, &c.; their laying out and management, and of the pruning and training of trees in different forms. The *fourth* contains abridged descriptions of the best fruits, a chapter on gathering and preserving fruits, another on diseases and insects, and another on the implements in common use.

Illustrations have been introduced wherever the nature of the subject seemed to require them, and it was possible to get them prepared. It is believed that these will prove of great value in imparting a correct knowledge of the various subjects. Upwards of one hundred of the more important figures have been drawn from nature by Prof. Sintzenich of Rochester.

P. B.

Mount Hope Garden and Nurseries,
Rochester N. Y.

CONTENTS.

PART I.

GENERAL PRINCIPLES.

CHAPTER I.
NAMES, DESCRIPTIONS, AND OFFICES OF THE DIFFERENT PARTS OF FRUIT TREES, ... 1

CHAPTER II.
SOILS, ... 48

CHAPTER III.
MANURES, ... 54

CHAPTER IV.
THE DIFFERENT MODES OF PROPAGATING FRUIT TREES, . . 60

CHAPTER V.
PRUNING—ITS PRINCIPLES AND PRACTICE, . . . 83

PART II.

THE NURSERY 105

PART III.
PLANTATIONS, ORCHARDS, ETC.

CHAPTER I.
PERMANENT PLANTATIONS OF FRUIT TREES, . . . 157

CHAPTER II.
PRUNING APPLIED TO THE DIFFERENT SPECIES OF FRUIT TREES UNDER DIFFERENT FORMS, ... 203

PART IV.
SELECT VARIETIES OF FRUITS.

CHAPTER I.
ABRIDGED DESCRIPTIONS OF SELECT VARIETIES OF FRUITS, . 277

CHAPTER II.
GATHERING, PACKING, TRANSPORTATION AND PRESERVATION OF FRUITS, 354

CHAPTER III.
DISEASES AND INSECTS, ... 36?

CHAPTER IV.
NURSERY, ORCHARD AND FRUIT GARDEN IMPLEMENTS . . 377

PART I.

GENERAL PRINCIPLES.

GENERAL PRINCIPLES.

CHAPTER I.

NAMES, DESCRIPTIONS, AND OFFICES OF THE DIFFERENT PARTS OF FRUIT TREES.

General Remarks.—A Tree is a living body composed of many parts, such as roots, branches, leaves, buds, blossoms, fruit, &c. All these have different offices to fulfil, assume different forms and characters, and are known and designated from one another by different names when subjected to the practical operation of culture. Without some knowledge of the names and structure of these different parts, of the principles that guide their development, their relative connection with, and influence upon one another, tree culture cannot be, to any man, really pleasant, intellectual, or successful; but a misty, uncertain, unintelligible routine of manual labor.

The industry of our times is peculiarly distinguished by the application of science—the union of theory with practice in every department; and surely the votaries of the garden, whose labors, of

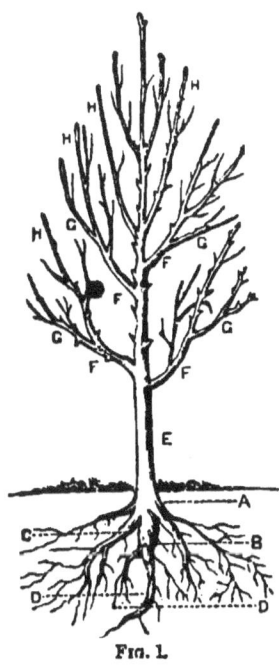

Fig. 1.

Fig. 1, a tree. *A*, the collar. *B*, the main root. *C*, lateral root. *D*, fibres. *E*, stem or trunk. *F*, main branches. *G*, secondary branches. *H*, shoots, one year's growth.

1

all others, should be intelligent, will not allow themselves to fall behind and perform their labors in the dark.

Fully sensible of the importance of this preliminary study, and confident that the minute and practical details of culture cannot be well understood without it, I propose here, before entering upon the main subject, to describe, in as few and as plain words as possible, the structure, character, connection, and respective offices of the various parts of fruit trees, and the names by which each is known in practice.

SECTION 1.—THE ROOT.

THE ROOT is composed of several parts.

1st. The *collar* (*A*, fig. 1), which is the centre of growth, or point of union between the root and stem, usually at or just below the surface of the ground. In root grafting seedlings, this is the point where the graft is set.

2d. *The body or main root* (*B*, fig. 1), which usually penetrates the earth in a *vertical* direction, and decreases in size as it proceeds downwards from the collar. It is also called the *tap root*. A seedling that has not been transplanted has usually but one descending or tap root, furnished in all its length with minute hairy fibres.

3d. *The lateral roots* (*C*, fig. 1) are principal divisions or branches of the main root, and take more or less of a spreading or horizontal direction. When seedlings are transplanted, having a portion of the tap root cut off, these lateral or side roots are immediately formed.

4th. *The fibres or rootlets* (*D*, fig. 1) are the minute hair-like roots which we see most abundant on trees that have been frequently transplanted. Different species of trees vary much in their natural tendency to produce fibres. Thus the pear and the apple require frequent transplanting, and often root pruning, to produce that fibrous condition which is necessary to great fruitfulness; whilst the paradise apple, used as a stock for dwarf trees, and the quince, are always quite fibrous, the former never, and the latter seldom requiring root pruning.

5th. *The Spongioles* are the extremities of the fibres, porous and spongy, through which the food of trees derived from the soil is mainly absorbed; these points are composed of soft, newly formed, delicate tissue, and are exceedingly susceptible of injury. The slightest bruise or exposure to a dry or cold air is fatal to them; and this is the reason why transplanted trees receive generally such a severe check and so frequently die. If trees could be taken up in such a way that these spongioles could all, or mostly, be preserved, trees would receive no check whatever; hence large trees are removed in midsummer without a leaf flagging.

6th. *Growth of Roots.*—The most popular theory at this time is—that the growth of roots is produced by the prolongment of the woody vessels of the stem, which descend in successive layers to the extremities of the roots, and thus promote their extension.

When these descending layers are interrupted in their course by some natural or accidental cause, or by art, as when we cut off the ends of roots, they pierce the bark and form new roots or new divisions of the root in the same manner that branches are produced on the stem. Thus the roots furnish food to the stem and branches for their support and enlargement, and in return, the stem and branches send down layers of young wood to increase and solidify the root; the one depending entirely upon the other for its growth and existence. Practical cultivators are familiar with many facts that illustrate the intimate relations and mutual dependency of the roots and stems. For instance, where one portion of the head or branches is much larger or more vigorous than the other, if the roots be examined, it will be found that those immediately under, or in direct connection with the largest branches, will have a corresponding size and vigor. In cases where one side of the top of a large tree is cut off, as in top grafting, a large number of new shoots are produced on the cut branch, and, if the roots be examined under or in connection with this branch, a corresponding new growth will be found there. It is quite obvious from these and similar facts, that whatever affects the roots or stems of trees favorably or unfavorably, affects the whole tree. If the foliage of a tree be entirely removed in the growing season, the absorbent action of the roots is suspended; and if the spongioles or absorbing points of the roots be cut off, the growth of the top instantly ceases. Those who have leisure should pursue the study of these highly interesting and important points still further.

SECTION 2.—THE STEM.

The Stem is that part of a tree which starts from the collar and grows upwards. It sustains all the branches, and forms the channel of communication between the different parts of the tree from one extremity to the other.

Plants like the grape, with twining or climbing stems, are called *vines*, and such as have no main stem, but have branches diverging from the collar, as the gooseberry, currant, &c., are called *shrubs* or *bushes*. Where the stem is destitute of branches to some distance from the ground, it is usually called the *trunk*.

Different Parts of the Stem.—A stem or branch of a tree is composed of the following parts, which are distinctly observable when we cut it across. Fig. 2 represents the half of a cross section of the stem of a young tree five years old.

1. *The Rind or Outer Bark* (*A*) on shoots or young parts of trees; this is thin, smooth, and delicate, like tissue

paper, and is easily separated from the parts beneath it.

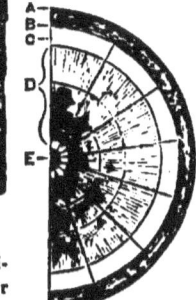

FIG. 2 F.

Fig. 2, half of the horizontal section of a five year old stem. A, outer bark or rind. B, inner bark or liber. C, sap-wood or last formed layer. D, perfect wood. E, pith. F, vertical section of a five year old stem, showing the five layers of perpendicular woody tubes or fibres.

FIG. 2.

In some species, as the grape vine, for example, this rind is shed and renewed annually, whilst in others, as the apple, pear, &c., it unites with the layer of tissue beneath it, and forms a hard, scaly, or corky substance, usually called *cortical layers*, which separate from the tree at different periods, according to the age of the subject and other circumstances.

It is these cortical layers that give rise to the expressions *smooth* and *rough* back.

2. *The Inner Bark or Liber* (B.)—This is the interior portion of the bark in immediate contact with the wood. It is composed of perpendicular layers of soft, flexible fibres, filled up with tissue. It is this part of the bark of the Basswood that is used for budding ties, &c., the tissue being separated from the woody fibre by maceration.

3. *The Sap-wood* (C.)—This is the youngest or last-formed layer of wood, immediately below the inner bark. It is distinguished in all trees by being softer and lighter colored than the older parts.

4. *The Heart or Perfect-wood* (D.)—This is the central or interior portion of the stem or branch, grown firm and mature by age. It is generally a shade darker in color than the newly-formed part or sap-wood.

5. *The Pith* (E.)—This is the soft, spongy substance in the centre of the stem and branches. In soft-wooded species, like the grape vine, it is large; in hard-wooded

species, as the apple, pear, quince, &c., small. In young shoots it is soft, green, and succulent, and fills an important part in their development. In the old part it is dry, shrivelled, and seems incapable of taking any part in the process of vegetation, and this appears evident from the fact that trees often continue to flourish after the centre, containing the pith, has begun to decay.

Structure of the Stem.—The stem is composed of woody fibre and cellular tissue, a substance similar to the pith. The woody fibre is arranged in perpendicular layers, and the cellular tissue in horizontal layers, running from the pith to the bark and connecting them. The mingling of these two systems gives to the surface of the cross section of a stem the beautiful veined or netted appearance observable in fig. 2. The perpendicular layers of woody fibre are most clearly observable when we cut a stem vertically; they are then easily separated from one another. The layers or plates of tissue radiating from the centre to the stem are usually called the *medullary rays.*

The inner bark or liber, as has been stated, is, like the wood, composed of thin layers of delicate perpendicular fibres mixed with tissue.

Growth of the Stem.—The stem of a tree is originally the extension of the cellular tissue of the seed. As soon as leaves are formed they organize new matter, which descends and forms woody fibres: the layers sent down from the first leaves are covered with those sent down from the next, and so on, one layer after another is produced until the end of the season, when the leaves fall and growth ceases. A yearling tree has, therefore, a greater number of layers of woody fibre at the collar than at the top, and is, consequently, thicker; the second year the buds on the first year's growth produce shoots, and these organize new layers of woody fibre, that descend and cover those of the previous year, and thus growth proceeds from year to year. Between each year's growth there is generally a line, in some cases more conspicuous than in others, that marks off the formation of each year, so that we are able to reckon the ages of trees with great accuracy by these rings. When it happens that a tree, from certain circumstances, makes more growth one season than another, we find the ring of that season larger. The new wood is always formed between the inner bark and the last layer of wood, so that one layer is laid upon, and *outside* of another, and the bark is continually pressed outwards.

The new layers of bark are also formed at the same place, or *within* the previous one. From this mode of growth, it results that each layer of wood is more deeply imbedded as others are formed on the top of it; and each layer of bark is pressed outwards as others are formed within it. In some cases, as in the cherry, for example, the bark is so tough as not always to yield to the general expansion of the tree, and slitting is resorted to for the purpose of preventing an unnatural rupture, which would eventually take place by the continued pressure of growth from within.

SECTION 3.—BRANCHES.

Branches are the divisions of the stem, and have an organization precisely similar : they are designated as,

1st. *Main Branches* (F, fig. 1) ; those that are directly connected with the stem or trunk. In pyramidal trees, they are called *lateral branches*. The branches of different species and varieties of fruit trees, differ much in their habits of growth ; and it is highly important to the planter to consider these peculiarities, because certain habits of growth are better adapted to particular circumstances than others. Thus we have *erect branches* (fig. 3), which produce trees of an upright and compact form. *Curved erect branches* (fig. 4), proceeding almost horizontally from the stem for short distance, and then becoming erect; these, also,

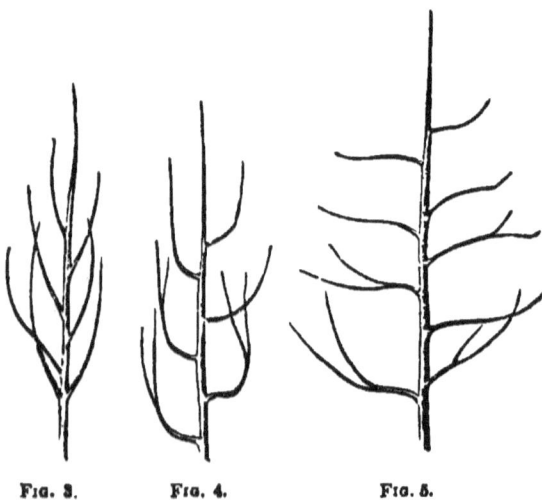

Fig. 3.   Fig. 4.   Fig. 5.

Figs. 3, 4, 5, different habits of growth of trees. 3, erect; 4, curved erect; 5, spreading or horizontal.

form upright symmetrical heads, but much more open than the preceding. Also, *horizontal* or *spreading branches* (fig. 5), that form wide-spreading heads with irregular outline. And, lastly, *drooping branches*, when they fall below the horizontal line. The branches of most varieties of apples and pears become pendulous when they have borne some time; and even in young trees of particular varieties, some of the branches assume a drooping and irregular habit.

2d. *Secondary Branches* (*G*, fig. 1), are the divisions of the main branches: occasionally those near the stem take such a prominent part in forming the outline of the tree, as to assume all the character of main branches, excepting in position.

3d. *Shoots* (*II*, fig. 1). This is the name by

young parts are designated from the time they emerge from the bud until they have completed their first season's growth. These have also important peculiarities that serve to distinguish certain varieties. They are variously designated as *stout* or *slender*, *stiff* or *flexible*, *erect* or *spreading*, *short jointed* if the buds be close together, and *long jointed* when the contrary. The *colors* of their bark are also strikingly different, and form very obvious distinctions amongst varieties. The *Snow Peach*, for instance, has pale greenish shoots, by which it is at once distinguished. The *Jargonelle, Rostiezer*, and many other var'eties of the pear, have *dark purplish* shoots, while the *Dix* and *St. Germain* are quite *yellowish*, the *Glout*

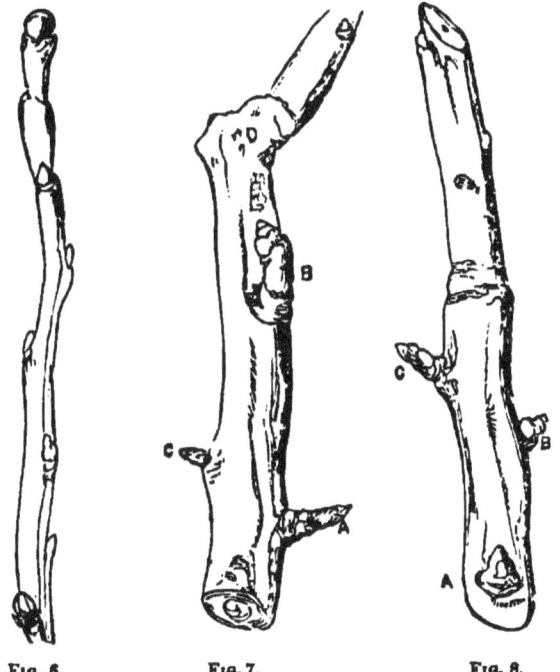

Fig. 6.   Fig. 7.   Fig. 8.

Fig 6, wood branch of the apple. 7, fruit branch; *A, B, C,* young spurs on two year-old wood. 8, fruit branch of the pear; *A, B, C,* young spurs on two-year-old wood.

Morceau, *grey* or *drab*, and the *Bartlett* and *Buffam* quite *reddish*. The shoots of certain varieties of apples and pears, and especially plums, are distinguished by being *downy*, furnished to a greater or less extent with a soft and hairy covering—in some cases barely observable.

4th. *Wood Branches* (fig. 6) are those bearing only wood buds.

5th. *Fruit Branches* are those bearing fruit buds exclusively. They are presented to us under different forms and circumstances, all of which it is of the highest importance to understand.

In *kernel fruits*, such as the apple and pear, the most ordinary form of the fruit branch is that generally called the *fruit spur* (*A*, *B*, *C*, figs. 7, 8, 9). It appears first as a prominent bud, as in fig. 7, on wood at least two years old; and for two or three seasons it produces but a rosette of leaves, and continues to increase in length, as in fig. 9. After it has produced fruit, it generally branches, and, if properly managed, will bear fruit for many years. Apple and pear trees of bearing age, and in a fruitful condition, will be found covered with these spurs on all parts of the head except the young shoots. In addition to the *fruit spur*, there are on the kernel fruits slender *fruit branches*, about as large as a goose quill, and from six to eight inches in length (fig. 10); the buds are long,

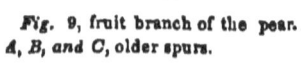

Fig. 9.

Fig. 9, fruit branch of the pear. A, B, and C, older spurs.

narrow, and prominent, and the first year or two after their appearance, produce but rosettes of leaves, yielding fruit generally about the third year. On trees well furnished with fruit spurs, these slender branches are of little account, but they are useful on young trees not fully in a bearing state. They are generally produced on the lower or older parts of the branches or stem, and, in the first place, are slender shoots with wood buds only; but owing to their unfavorable position and feeble structure, they receive only a small portion of the ascending sap, and the consequence is, they become stunted, and transformed into fruit branches. In pruning young trees, slender shoots are frequently bent over, or fastened in a crooked position to transform them into fruit branches of this kind; but this will be treated of in its proper place.

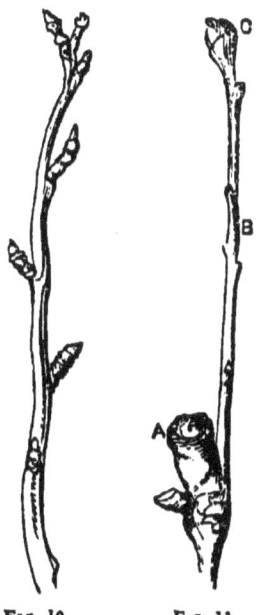

Fig. 10. Fig. 11.

Fig. 10, slender fruit branch of the apple—all the buds are fruit buds. *Fig.* 11, a branch of the apple showing the tendency of some varieties to bear on the points of the branches. *A*, the point where a fruit was borne last season; *B*, a shoot of last year; *C*, its terminal fruit bud.

Certain varieties of apples have a natural habit of bearing the fruit on the points of the lateral shoots; and frequently these terminal fruit buds are formed during the first season's growth of the shoot. Fig. 11 is an example; *A* is the point where a fruit was borne last season; *B*, a shoot of last season; and *C* its terminal bud, which is a fruit bud. The fruit branches of the *peach*, *apricot*, and *nectarine*, are productions of one season's growth; the fruit buds form one season and blossom the next; but as on the

Fig. 12.

Fig. 12, fruit spur of the peach on the old wood.

Fig. 13

apple and pear, there are different forms of the fruit branch.

In the first place the *fruit spur* (*A*, fig. 12), a group of buds like a bouquet; these are little stunted branches on the older wood that have assumed this form. The most important fruit branches of these trees are the vigorous shoots of the last season's growth, containing both fruit and wood buds (fig. 13), and the slender fruit branches, bearing all single fruit buds, except a wood bud or two at the base. Fig. 14 represents such a branch of the peach, *A* and *B* being wood buds. The fruit branches of the *plum* and *cherry*, and the *gooseberry* and *currant*, are similarly produced. A yearling shoot, for instance, the second season, will produce a shoot from its terminal bud, and probably shoots from two or three other buds immediately below the terminal, whilst those lower down will be transformed into fruit buds, and produce fruit the third season. Fig. 15 is a branch of the cherry. *A* is the two-year-old wood; *B*, one year; *C* and *D*, fruit spurs on the two-year-old wood, with a wood bud usually at the point. Fig. 16 is a fruit spur from the older wood; *A*, the wood bud at its point.

Fig. 13, mixed wood and fruit branches of the peach; *C, D, E*, fruit buds; *F, G, H*, leaf buds; *I*, double buds; *C*, triple buds, the two side buds being fruit buds and the centre one a leaf bud.

Fig. 17 is a branch of the plum; *A*, the two-year-old wood; *B*, one year old; *C* and *D*, spurs. Fig. 18 is a

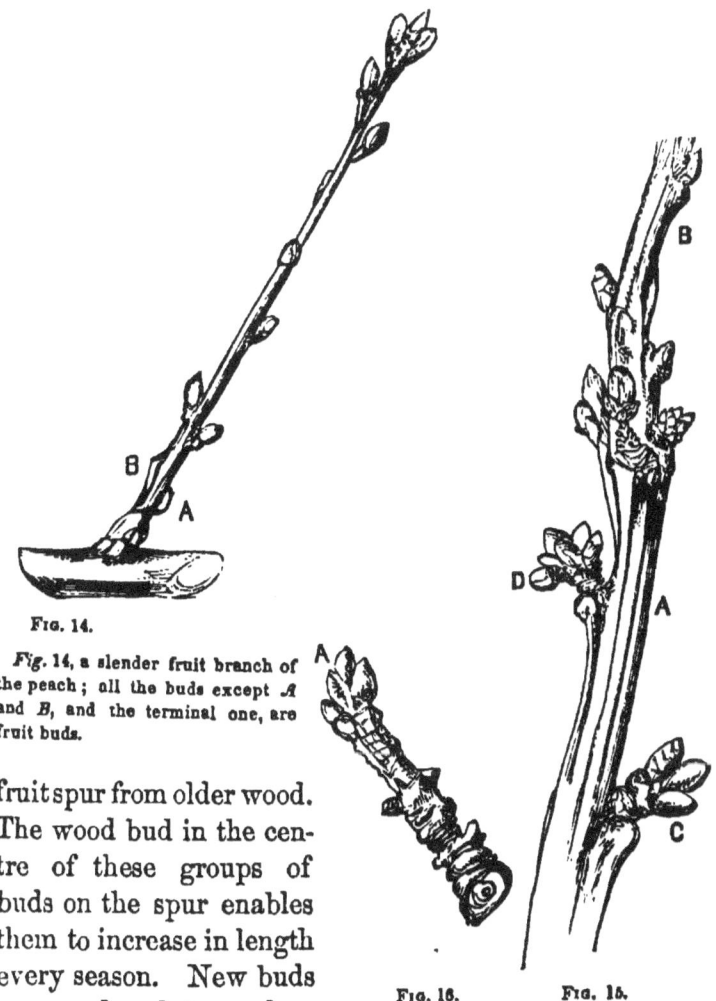

Fig. 14.

Fig. 14, a slender fruit branch of the peach; all the buds except *A* and *B*, and the terminal one, are fruit buds.

Fig. 16. Fig. 15.

Fig. 15, branch of the cherry; *A*, two-year-old wood; *B*, one year; *C* and *D*, fruit spurs. *Fig.* 16, fruit spur of the cherry; the bud *A*, in the centre of the group, is a wood bud.

fruit spur from older wood. The wood bud in the centre of these groups of buds on the spur enables them to increase in length every season. New buds are produced to replace those that bear, and so the spurs continue fruitful for several years, according to the vigor of the tree, and the manner in which it is treated.

The fruit branches of the *quince* and the *medlar* are slender twigs on the sides of lateral branches, and the fruit is borne on their points.

Fig. 18.   Fig. 17.

Fig. 17, branch of the plum; *A.* two-year-old wood; *B,* one year old; *C* and *D,* spurs. *Fig.* 18, fruit spur of the plum on the old wood.

Section 4.—Buds.

1st. *The Nature and Functions of Buds.*—In a practical point of view, buds are certainly the most important organs of trees, because it is through them we are enabled completely to direct and control their forms and their productiveness. Whoever, therefore, wishes to become a skilful and successful tree culturist, must not fail to make himself familiar with all their forms, modifications, modes of development, and the purposes they are adapted to fulfil in the formation of the tree and its products. The immediate causes of the production of buds on the growing shoots of trees, and the sources from which they spring or in which they originate, are alike thus far mysterious, notwithstanding they have been the subject of a vast deal of research and speculation among botanists and vegetable physiologists for many ages. We are able, however, to trace clearly and satisfactorily the objects they are intended to fulfil in the development of the tree, their connection with, and dependency upon other parts, and the circumstances under which they can be made to accomplish specific purposes.

Every bud contains the rudiments of, and is capable, under favorable circumstances, of producing a new individual similar to that on which it is borne.

This fact is clearly demonstrated in the propagation of trees by budding, where a single eye is removed from one shoot and placed in the wood of another, to which it unites and forms a new individual similar to its parent. So in propagation by eyes, as in the grape vine, where a single bud with a small portion of wood attached, becomes a perfect plant.

Every perfect bud we find on a young yearling tree or shoot is capable of being developed into a branch. Naturally, they do not; but we know that by the application of art they can be readily forced to do so.

For instance, the buds of a yearling tree, if left to take their natural course, will only in part produce branches, and these will generally be nearer to the extremities, where they are the most excitable, being in closer connection with the centre of vegetation: but we cause the lower ones to develope branches, by cutting off those above them to the extent that the particular character of the species or variety, or of the buds themselves in respect to vigor and vitality, may require. Hence it is that the forms of trees are so completely under our control when we possess the requisite knowledge of the character and modes of vegetation of buds.

2d. *Different Names and Characters of Buds.*—All buds are either, 1st, *terminal*, as when on the points of shoots (*C*, fig. 19); 2d, *axillary*, when accompanied by a leaf situated in the angle made by the projection of the leaf from the shoot or branch (*A B*, fig. 19); 3d, *adventi-*

tious or *accidental*, when originating accidentally as it were, or without any regularity, on the older parts of trees,

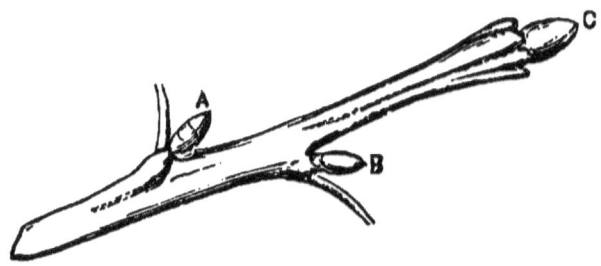

FIG. 19 —LATERAL BRANCH.

Fig. 19. *A*, a superior bud. *B*, inferior. *C*, terminal. *A* and *B*, axillary

and unaccompanied by a leaf. They are often produced by the breaking or cutting off a branch, or by a wound or incision made in the bark. In the management of trained trees special means are taken to produce these buds on spaces of the trunk that it is desirable to fill up. We sometimes see instances of such buds on the stumps of old trees.

The terminal and axillary buds produced on young shoots, seem to have a different origin from these accidental buds—the former are connected with the pith of the shoot, as we may see by dissecting them. On cutting into a young shoot below a bud we find a cylinder of pith entering into the bud from the pith of the shoot, but we do not find this connection existing in the case of the adventitious buds.

Practically considered, buds are classified as follows:—
1. *Lateral.*—Those on the sides or circumference of shoots, being the *axillary* buds of the botanist (*A*, *B*, 19).
2. *Terminal.*—Those on the points of shoots (*C*, fig. 19).
3. *Superior.*—Those on the upper sides of horizontal branches (*A*, fig. 19).

4. *Inferior.*—Those on the lower side of horizontal branches (*B*, fig. 19).
5. *Stipular.*—The small, barely visible buds found at the base of ordinary buds.
6. *Dormant* or *Latent.*—These are scarcely apparent buds, generally towards the base of branches: They may remain dormant for several years, and then, in some species, be excited into growth by pruning close to them.

Buds are again classed as *leaf buds* and *fruit buds*.

7. *Leaf Buds* (*F, G, H,* fig. 13) produce either leaves or branches; they differ in form from fruit buds in being in most cases longer and more pointed in the same species.

These are again designated as—

Single, when only one is produced at the same point (*H*, fig. 13).

Double, when two are together (*I*, fig. 13).

Triple, when in threes (*C* and *J*, fig. 13).

These double and triple buds are almost peculiar to the stone fruits, and especially the peach, apricot, and nectarine.

The size, form, and prominence of leaf buds vary in a striking degree in different varieties of the same species, and these peculiarities are found to be of considerable service in identifying and describing sorts. Thus, the buds of one variety will be long, pointed and compressed, or lying close to the shoot. Others will be large, oval and prominent, or standing boldly out from the shoot. Others will be small, full, and round. Thus, for instance, the wood buds of the *Glout Morceau* are short and conical, broad at the base, and taper suddenly to a very sharp point inclined towards the shoot; they have also very prominent shoulders, that is, their base forms a prominent projection on the shoot. The scales are also dark, with light gray edges. In the *Josephine de Maline* pear the buds are quite remarkable for their roundness, bluntness and prominence. If shoots of the *Bartlett* and *Seckel* pears, two well known varieties, be compared, although they present no decidedly obvious peculiarities, yet they will be found very different. Those of the *Seckel* are much broader at the base, more pointed, and lighter colored, being a dark *drab*, whilst those of the *Bartlett* are *reddish*. These miscellaneous instances are chosen simply to draw attention to these points, and to show the ordinary modes of comparison. When we speak of leaf buds, we have reference only to the simple bud and not to the large, pointed, spur-like productions frequently produced towards the middle or lower part of young shoots that have made a second growth, that is where growth has ceased for a while and the terminal bud has been formed, and afterwards, in the same season, commenced anew, and made a second growth.

8. *Fruit Buds.*—In the early stages of their formation and growth all buds are but leaf buds. Thus, on a young shoot of the cherry and the plum, for example, of one season's growth, the buds are all leaf buds. The next spring a part of these produce new shoots, and others are transformed into fruit buds that will bear fruit the following season. The transformation is accomplished during the second year of their existence, and it usually happens that they are the smallest and least fully developed that are so transformed: the more vigorous pushing into branches. In the peach, the apricot, &c., on which the fruit buds are produced in one year, the change from a leaf to a fruit bud occurs towards the latter part of the season. The primary cause of the transformation of leaf into fruit buds is not satisfactorily known, although many theories exist on the subject. Observation, however, has taught us many things in relation to it. It seems that all trees

must acquire a certain maturity, either natural or forced, in order to produce blossoms or fruit. A tree that is furnished with a rich, humid soil, containing an abundance of watery nutriment, and left in all respects unrestrained in its upward growth, may attain the age of ten or fifteen years before it commences to form fruit buds; whilst in a soil of a different quality, dry and less favorable to rapid growth, or if constrained in its growth by being grafted on some particular stock, or by some particular mode of training, it may produce fruit in two or three years.

An apple tree on a common stock, planted out in ordinary orchard soil, does not usually bear until it is in most cases seven years old from the bud, often more; whilst the same variety grafted or budded on a paradise apple stock will produce in two or three at most. We frequently see one branch of a tree that has been accidentally placed in a more horizontal position than the other parts, or that has been tightly compressed with a bandage or something of that sort, bear fruit abundantly; whilst the erect, unconstrained portion of the tree gives no sign of fruitfulness whatever. As a general thing, we find that where there is an abundant and constant supply of sap or nutriment furnished to the roots of trees and conveyed by them through the unrestrained channels which the large cells and porous character of young wood afford, the whole forces of the tree will be spent in the production of new shoots; but that as trees grow old, the cells become smaller, and the tree being also more branched the free course of the sap is obstructed, and becomes in consequence better elaborated, or in other words more *mature*, and commences the production of fruit. Circumstances similar in all respects to these and answering exactly the same purpose, can be produced by art at an early age of the tree; and this is one of the leading points in the culture and management of garden trees, where smallness of size and early fruitfulness are so highly desirable. This will come under consideration in another place.

Fruit buds in most cases are distinguishable from wood buds by their rounder and fuller form; the scales that cover them are broader and less numerous, and in the spring they begin to swell and show signs of opening at an earlier period. Like the wood buds they are *single*, *double*, or *triple*, according to the number found together. They are *single* in pears, apples, and other trees of that class. *Single*, *double, and triple*, variously, on the stone fruits, gooseberries, and currants.

Fruit buds are also *simple* and *compound*. *Simple*, as in the *peach*, *apricot*, and *almond*, each bud of which produces but one flower. *Compound*, as in the *plum, cherry, apple, pear*, &c., each bud of which produces two or more flowers. Those of the plum produce two or three, hence we find plums usually borne in pairs; those of the cherry four or five (fig. 20), and of the apple and pear six to eight; and hence we often find these fruits borne in clusters. They are also *lateral* or *terminal*, as they occupy the sides or ends of the branches or spurs on which they are produced. The ordinary position of the fruit buds of different classes of trees will be understood from the preceding descriptions of fruit branches.

Fig. 20, flower of the cherry, showing the product of a compound bud.

Section 5.—Leaves.

1st. *Structure and Functions of Leaves*.—The leaves of all hardy fruit trees cultivated in our climate are deciduous, that is—they decay and fall in the autumn and are succeeded by others on the return of spring. The offices they perform during the growing season are of the highest importance to the life and health of the tree, and deserve the most attentive consideration.

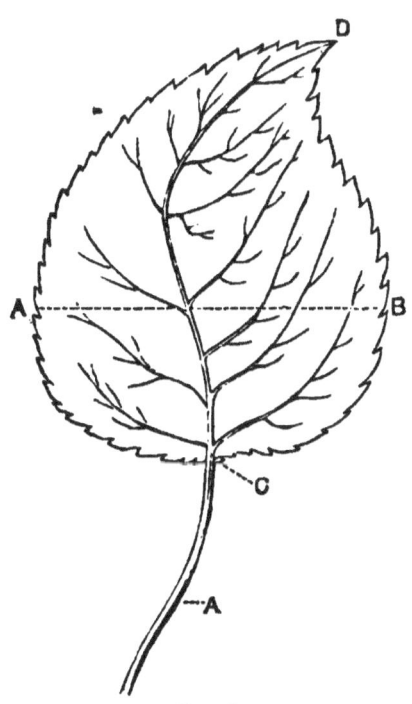

Fig. 21.

Fig. 21, a leaf of the pear. *A*, the petiole or leaf stalk. *A*, *B*, *D*, the blade. *C*, the base. *D*, the point. Line *A*, *B*, the width.

A leaf (fig. 21) is composed of two principal parts, the *leaf stalk* or *petiole* (*A*), which connects it with the tree or branch on which it is borne, and the expanded part (*A*, *B*, *D*), called the *blade*. The *base* is the end (*C*) attached to the stalk, and the *apex or point* (*D*) the opposite one. The *length* is the distance from the base to the point (*C* to *D*), and the *width*, a line cutting the length at right angles, and extending from margin to margin (*A* to *B*).

The leaf stalk and its branches, forming the nerves or veins of the blade, are composed of woody vessels in the form of a tube, similar to the woody parts of the tree or branch that bears it, inside of which is a pith, similar to the pith of the tree; the leaf is thus connected with the pith and wood of the shoot, and consequently the ascending sap, as we may readily see, by making a vertical cut through the leaf stalk and shoot. The veins of the leaf are filled up with a cellular substance similar to the pith, called *parenchyma*, and the whole is covered with a thin skin (epidermis). This cellular substance is connected with the inner bark, and consequently the descending sap or cambium, that forms the new layers of wood. Both surfaces of the leaf are furnished with small pores, through which exhalation and absorption are carried on. Absorption is performed principally by the pores of the under surface, and they are the largest; exhalation principally by those of the upper surface.

This property of the leaves to receive and give out air and moisture through the pores on their surface, has caused them to be likened to the lungs of animals, and this comparison is to some extent correct; for we know that without leaves, or organs performing their offices, trees do not grow. And in proportion to their natural and healthy action, do we find the vigor and growth of the tree.

To prove that leaves have the power, in a greater or less degree, to absorb fluids, we have but to apply water to the drooping foliage of a plant suffering from drought, and see how quickly it becomes refreshed. Dews of a single night, we know, too, will revive plants that the heat and drought of the previous day had prostrated; and even if we put a flagging plant in a damp atmosphere, it recovers. Even the leaves of a boquet can be kept fresh for a long time by sprinkling them with water.

That plants *exhale* moisture and gases cannot be doubted. It is this very exhaling process that causes plants to wilt under a hot sun or in a dry atmosphere. Plants that are transplanted with their foliage on, as annuals are in the spring or summer, will wilt and even die if exposed to the air and sun; but if transplanted in a moist day, or covered, so that evaporation cannot take place, the plant does not appear to feel the removal. So with cuttings of many plants thus propagated; if placed in the earth with a certain amount of foliage on, and left uncovered, they will immediately die; but when we place a bell glass or a hand glass over them to prevent evaporation, they remain as fresh as though they had roots supplying them with moisture from the soil. It is on this account that transplanted trees so often die when the branches and shoots are not in proportion to the roots. In transplanting, a portion of the roots are destroyed, and all are more or less deranged, so that their functions are feebly performed for some time after planting. If all the branches and shoots are left on, they will, as usual, produce leaves, but the absorption at the roots being so much less than the exhalation of the leaves, the juices contained in the tree, previously laid up, soon become exhausted, the leaves droop and wither, and the whole fabric perishes. In budding, too, if the whole leaf were left attached, the evaporation would be so great as to kill the bud; hence we remove all but a portion of the stalk.

A tree can neither mature its wood nor its fruit without the full and healthy exercise of the leaves. If in the growing season, a tree is deprived of its foliage by blight, insects, &c., we see that growth is entirely suspended for a time, until new leaves are developed; and if the leaves be removed from a tree bearing fruit, we see the fruit shrivel and dry up, or ripen prematurely and become worthless. These facts, and many others that might be cited, show the intimate connection existing between the leaves and the other organs of trees, and the influence

they have on their growth and productiveness. It is believed that the opening of the leaf buds in spring induces the formation of new roots; this is doubtful, as new roots may be seen forming at times when there are no leaves on the tree and apparently no growth whatever going on in the buds. But if the roots are not roused into action by the leaves, it is well known they will not continue and grow long if leaves do not make their appearance. We observe in the case of trees the tops of which have been so much injured by drying and exposure that scarcely a sound bud is left to grow, in this case the roots, although in perfect order, remain nearly dormant until new shoots and leaves are produced, and in proportion as the leaves increase so do the roots. The fact of the absorption and exhalation by leaves of certain fluids, has, to a very considerable extent, established the theory that the sap of trees is taken up from the roots through the cells or sap vessels of the wood of the trunk and branches in a *crude state*, and passes into the leaves; that in their tissue spread out under the sun's rays, it receives certain modifications. Carbonic acid, which has been taken in a state of solution from the soil and by the leaves from the atmosphere, is decomposed, its oxygen is given off into the air, carbon becomes fixed, and thus the component parts of the tree, starch, sugar, gum, &c., are formed. After passing through this purifying or concentrating process, the sap acquires a more solid consistence, and is called *cambium;* so prepared it returns downwards through the nerves or vessels of the leaf to the base of the leaf stalk, and then between the wood and bark of the stem, forming new layers on its passage. Such is, at present, the most popular theory of the functions of the leaves and the ascent, assimilation, and descent of the sap. Some distinguished writers on the subject reject this theory, alleging that—
" there is no such thing as crude sap, that as soon as it

enters the roots it becomes assimilated and fit for the production of new cells, and that it passes upwards, forming new wood or cells by a chemical process."* Observation, however, has clearly established that in the leaves of healthy trees chemical processes depending on light and heat, and absolutely essential to the well-being of the tree, are continually going on, for trees shut out from the light always make a feeble growth and have a blanched and sickly hue, compared with the same species in the free air and exposed to the rays of the sun. If one side or portion of a tree is shaded or deprived of its full share of light, it ceases to grow in its natural way, and the shoots are lean, slender, and imperfect.

2d. *Different Forms and Characters of Leaves.*—The different sizes and forms of the leaves of fruit trees, the divisions of their edges, the absence or presence of glands, the smoothness or roughness of their surfaces, are all more or less serviceable in describing and identifying varieties.

The terms designating forms are seldom mathematically correct, but merely made by comparison, for instance—

Oval (fig. 22), when about twice as long as broad, and nearly of equal width at both ends.

Oblong (fig. 23), three times or more, as long as broad, and differing but little in width in any part.

Lance Shaped (fig. 24), lanceolate, when three or more times as long as broad, and tapering gradually to a sharp point.

Ovate (fig. 25), when twice as long as broad, tapering to the apex, and widest towards the base.

Obovate (fig. 26), the inverse of ovate, the greatest diameter being in the upper part.

* Schleiden's Principles of Botany

Round, roundish, as they approach a circular form like fig. 21. The point is often a distinguishing feature, some terminating suddenly in a sharp point, others drawn out to a long, sharp point, *peaked*, whilst others are nearly round. They differ much, too, in the form of the base, some are rounded, some sharp, and some heart-shaped.

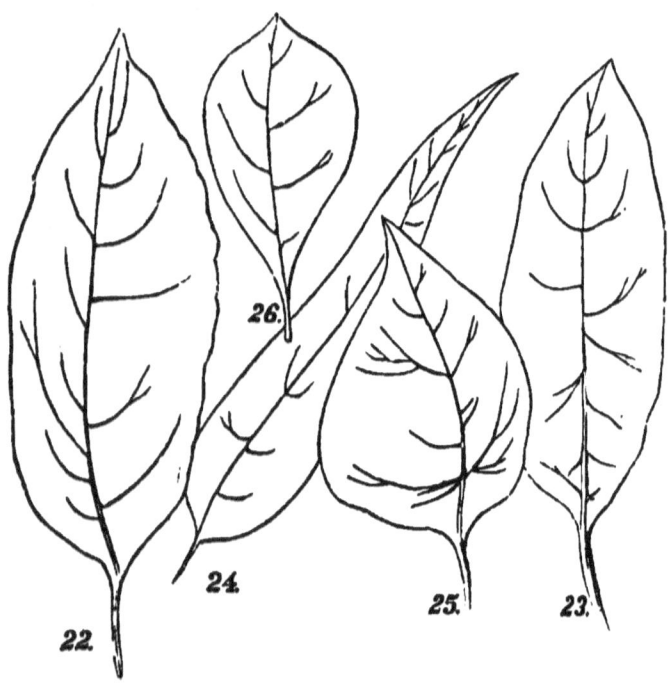

Figs. 22 to 26, forms of leaves. 22, oval. 23, oblong. 24, lanceolate. 25, ovate. 26, obovate.

The divisions of the edges are *serrated* or *toothed*, when the edges are cut into sharp teeth, directed towards the point of the leaf; *finely* (fig. 27) or *coarsely* (fig. 28) *serrate*, as these teeth are fine or coarse; *doubly serrate*, when the principal division or tooth is subdivided.

Crenate (fig. 29), when the divisions are rounded, instead of being sharp like teeth.

Lobed, when deeply cut, and the penetrating angle large, as in the currant, gooseberry, grape, &c. (Fig. 30).

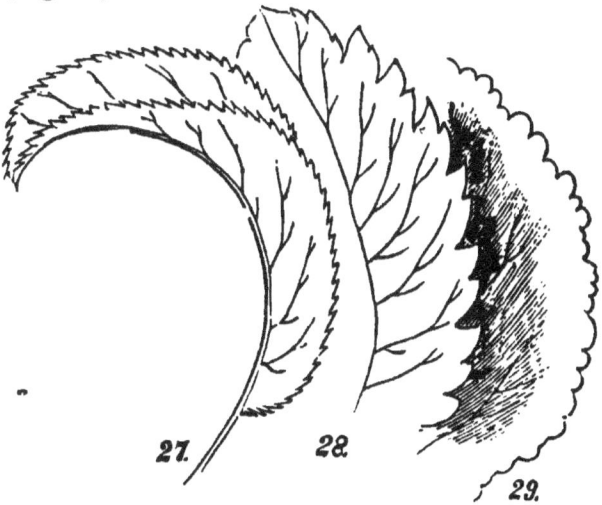

Fig. 27, a leaf, folded, reflexed, and finely serrated or toothed. Fig. 28, coarsely serrated. Fig. 29, crenate.

Fig. 30, a leaf of the currant, lobed.

Flat, when the surface is even (fig. 21).
Folded, when the edges are turned inward (fig. 27).
Reflexed, when the apex or point turns backwards, giving the leaf more or less the form of a ring (27).
Waved, wrinkled, smooth, rough;

etc., are all terms used, but well enough understood.

The leaf stalk has often striking peculiarities in certain varieties, such as unusually *long, stout, short,* or *slender*. There are also *glands* on the leaf-stalk, close to the base, and in certain cases on the leaf itself, that are chiefly taken notice of in identifying varieties of the peach and nectarine; these differ in shape too, being *globular* (as in fig. 31), reniform or *kidney-shaped* (fig. 32); these little glands are supposed to be, and no doubt are, organs of secretion.

Fig. 31.

Fig. 31, a leaf of the peach with globular glands.

Fig. 32.

Fig. 32, the same; with reniform or kidney-shaped glands.

These are all interesting items in the study of the beautiful and almost endless variety of forms which the different classes of fruit trees, and even different varieties of the same class, exhibit in their foliage.

Section 6.—Flowers.

1st. *Different Parts of Flowers.*—Flowers are the principal reproductive organs of trees, and consist of floral envelopes, the *calyx* and *corolla;* and of sexual organs, *stamens* and *pistils*.

The *Calyx* (A, fig. 35) is the outer covering, and is usually green like the leaves. The *corolla* (A, fig. 33) is within the *calyx*, and is the colored, showy part of the flower; its divisions are called *petals*.

Stamens (fig. 34) are the male organs of plants. They are delicate, thread-like productions (A, fig. 34) in the centre of the flower, supporting on their extremities the anthers

(*3*, fig. 34). The *pistil* (*C, D*, fig. 35) is the female organ and stands in the centre of the stamens. It consists of the *ovary* at its base (*B*, fig. 35), which contains the seeds. The *style* (*C*, fig. 35) is the erect portion, and the *stigma* (*D*, fig. 35) is the small glandulous body on its summit that receives the fertilizing powder (pollen) (*C*, fig. 34) from the anthers.

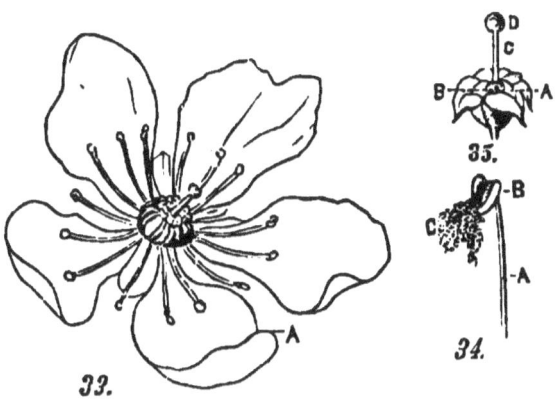

Figs. 33 to 35—Different parts of a flower. Fig. 33, *A*, the petals. 34, a stamen, *A*, filament or stalk. *B*, anther. *C*, pollen. 35, the calyx, ovary, pistil united. *B*, ovary. *C*, style. *D*, stigma.

Flowers may be deficient in any of these organs except the *ovary*, *anthers*, and *stigma*. These are indispensable to fructification, and must be present in some form or other or the flowers will be barren.

2d. *Sexual Distinctions.*—The fact that the two sexes or sexual organs, the *stamens* and *pistils*, are in certain species united on the same flower, and in others on different flowers, and even on different trees, has created the necessity for the following distinctions:

Trees or plants are called *hermaphrodite* (as in fig. 33) when both *stamens* and *pistils* are present on the same flower. Nearly all our cultivated fruits are of this class. *Monœcious*, when the male and female flowers are borne on the same tree, as in the filbert flower (fig. 36, *A*, the male, and *B*, the female flowers). *Diœcious*, when the male flowers (fig. 37) are on one plant, and the female (fig. 38) on another. The most familiar instance among plants cultivated for their fruits, is the *strawberry*. In many varieties we find the stamens or male organs so incompletely developed (fig. 38) that they are of no service in fructifying the flowers, and hence we plant near them varieties with an abundance of these organs strongly exhibited.

Fig. 37.

Fig. 37, male or staminate flower of the strawberry.

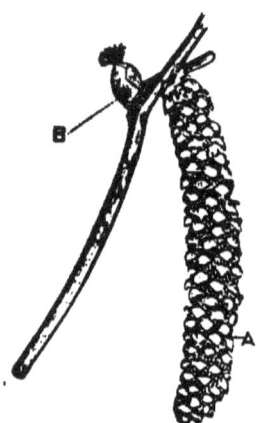

Fig. 36.

Fig. 36, flower of the filbert.

3d. *Impregnation.—* The process of impregnation is effected in this way: When the flowers first open, the pollen granules or powder in the anthers, is covered over by a delicate membrane. In a short time this membrane bursts in a manner similar to an explosion that scatters the pollen by its force, so that it reaches the stigma of the pistil; this is composed of glutinous or sticky secretions to which the pollen adheres; there it forms new cells that expand into tubes; these tubes penetrate through the style of the pistil to the ovary, where the impregnation takes place, and new cells are immediately formed into an embryo plant.

Fig. 38

Fig. 38, female or pistillate flower.

This impregnation is sometimes, from certain causes, only partially effected in the cases of fruit where the ovary or seed vessel is composed of several cells, as in the apple, pear, &c., and hence the fruit takes an imperfect, one-sided development from the beginning.

The difficulty that appears to arise in the way of the impregnation of the stigma of one flower by the pollen of another, distantly situated, either on the same plant as in monœcious trees, or on a different plant as in diœcious, is wonderfully obviated by the provision that nature has made for its transmission, not only by the atmosphere, but by insects, that pass from one flower to another feeding on their honied secretions; the pollen adheres to them, and they carry it from one to another.

All natural flowers of the same species present the same number of petals in their flowers, but occasionally the *stamens* are converted into petals, and thus what are called *double* flowers are produced. Among fruit trees we have double flowering apples, plums, peaches, and cherries. These seldom produce fruit; when perfectly double *never*. All our double flowers, roses, paeonies, dahlias, &c., have been obtained by this transformation of the stamens into petals. It is supposed to be caused by an excessively high cultivation given to the plants that produce the seeds from which these double varieties spring.

4. *Period of Blossoming.*—In treating of fruit buds allusion has been made to the causes which, according to observation and experience, promote fruitfulness. These are chiefly a slow or moderate growth, and a branching or spreading, constrained form, instead of an upright one. Some species of trees bloom at a much earlier age than others. Thus the peach, the apricot, and the cherry will bloom in nearly one-fourth less time from the bud, all things being equal, than the pear. Some species bloom at an earlier period of the season than others; the apricot and the peach bloom very early, and this is the chief reason why the crop is so often destroyed in localities subject to late spring frosts. Among fruits even of the same species there is much difference in the period of blooming: one variety of apple being nearly two weeks later than another. This, in some sections, is an important quality, where every day the blossom is retarded renders the crop

surer, being more likely to escape frost. These differences are caused by various circumstances.

1st. *The Climate.*—The period of blossoming of the same species varies much in different localities. Rochester is at least a week earlier than Buffalo, although the distance is less than one hundred miles; and it is nearly two weeks earlier than Toronto, which is still nearer. The large bodies of ice in the lakes, at both Buffalo and Toronto, have no doubt a considerable effect in retarding the blossoming period.

2d. *The Season and Position.*—In the same locality, one season is frequently a week earlier than others, and trees on the south side of a wall or building will expand their blossoms several days before the same variety in the open ground only a few rods distant, and ten days to a fortnight before those on a north wall.

3d. *The Soil.*—On warm and light soils, the roots of trees are excited into activity much sooner than in cold, damp, and heavy soils, and the blossoming period is earlier in consequence.

The Different Character of Flowers.—Flowers vary in *size, form, color,* and other qualities, even in the same species. In the peach those distinctions are so obvious, that one of the principal classifications of pomologists is founded on them. Thus there are varieties with *large, showy flowers* (fig. 39), as the *serrate early York,* and small (fig. 40) as *large early York, Crawford Early,* etc. The color also presents variations, some being *deep,* others *pale* rose, and some *almost white;* two or three varieties of the peach have flowers wholly white, as the *snow,* for instance. In all the other fruits,

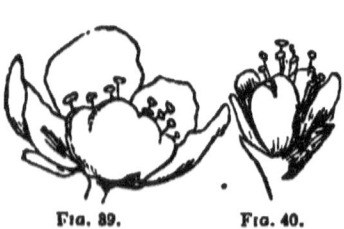

Fig. 39. Fig. 40.
Fig. 39, large flower of the peach.
Fig. 40, small flower of the peach.

as in *apples, pears, plums, cherries*, &c., the flowers vary but slightly in form and color, and the differences are only taken note of in very full and minute scientific descriptions. A few cases, however, are well marked, as the *Jargonelle* pear, the flowers of which are nearly twice as large as most others.

In connection with the flowers it may be proper to explain the important process of

Hybridization.—This is performed by fertilizing the pistil of one species or variety with pollen from the stamens of another. The seeds produced by the flower so impregnated will produce a cross or hybrid between the two parents. This process is now well understood, and is carried on to a wonderful extent, especially in the production of new flowers. Comparatively few of our popular fruits have been produced in this way. A few good sorts have been produced by the late Mr. Knight, a distinguished English experimentalist, who effected much in his time towards establishing many difficult and disputed points in vegetable physiology. Nearly all the native fruits of this country are accidental hybrids. A vast deal may be done to improve, in this way, all our fruits. The size, hardiness, and productiveness of one variety may be combined with the delicacy of texture and flavor of another, and endless variations and improvements may be effected. To obtain a true hybrid certain precautions are necessary. The two subjects selected must flower at the same time. The stamens must be carefully removed from the one intended for the mother, without injury to the stigma. It must also be guarded from accidental impregnation by other varieties, and the pollen from the selected male be applied at the proper moment, that is, when it bursts from the anther. Hybridization is only possible between species *closely related*, for although there is a relation between the *apple* and the *pear*, and between the *gooseberry* and the *currant*, they will not hybridize; but different *varieties* of the apple will hybridize with each other, and so with all the rest.

Blossoming in Alternate Years.—Many varieties of apple, pears, &c., fruits that take the whole season to mature, produce flowers in alternate years only, with great regularity. The reason is supposed to be this: The fruit during the bearing year, attracts a large quantity of the ascending sap of the tree in the same way as the leaves do; but instead of returning it to the tree, they consume it themselves. The consequence is, the buds that would have blossomed the following year if they had received their due share of nutriment, fail in attaining the proper condition, and produce only rosettes of leaves. During the unfruitful season, immense quantities of fruit-buds are again brought forward, and the year following, the tree is overloaded; so it proceeds in regular succession.

This is never experienced in trees regularly pruned, and may be remedied by thinning out the crop in bearing years, leaving on but a reasonable amount that will not exhaust the tree. The bearing years have been completely reversed by removing the blossom-buds or fruits on the bearing year.

Section 7.—The Fruit.

1st. *Character of the Fruit.*—As soon as the ovary is impregnated it begins to swell; the petals, stamens, and other parts of the flower fall off, and we then say the fruit is "*set.*" As a fruit bud is but a transformed leaf-bud, a fruit occupies the same relative connection with the tree as a branch; it attracts food from the stem and the atmosphere in the same manner, and performs all the same functions, except that it does not, like the leaf.

return anything to the tree, but appropriates all to its own use; and this is the reason, as we have before remarked, that trees having borne a heavy crop of fruit one season are unfruitful the next—this is the case only with fruits, as the apple and pear, that require nearly the whole season to mature them. Cherries, and other fruits that mature in a shorter period, and that draw more lightly on the juices of the tree, do not produce this exhaustion, and consequently bear year after year uninterruptedly.

2d. *Classification.*—In some fruits, as the apple for instance, the fruit is formed *below* or at the base of the calyx, the segments of which are still visible in the mature fruit; and often serves to some extent by its size and other peculiarities, as being spread out, or closed together in a point, to identify varieties. In other species, as the plum and cherry, the fruit is formed *within* the calyx, or on the top of it. Fruits of the former character forming below the calyx and including it in their structure are classed as *inferior*—the *apple, pear, quince, gooseberry,* and *currant* are all inferior, having the calyx adhering.

Those formed within the calyx, having the pistil alone connected with the ovary, are called *superior;* such are the *peach, plum, apricot, nectarine, cherry, raspberry, strawberry,* and *grape.*

The more natural, popular, and useful classification of fruits, is that by which they are divided into

Pomes or Kernel Fruits, as the *apple, pear, quince, medlar,* etc. In speaking of these we call the pericarp the *flesh,* and the dry, bony seed capsules the *core.*

Drupes or Stone Fruits.—Those having a soft, pulpy pericarp, and the seed enclosed in a shell like a nut, as the *peach, plum, apricot, cherry,* etc. The pericarp of these is called the *flesh,* and the seed, the pit or stone

Berries.—These have soft, pulpy flesh, containing seeds without capsules, as the *gooseberry, currant, raspberry, strawberry*, and *grape*.

Nuts, or capsule fruit, as the *filbert, chestnut*, etc., the fruits of which are nuts contained in husks or cups, that when ripe, open and let the fruit drop.

The outlines or forms of fruits and their colors exhibit great variations, even in the same species. Every portion of the fruit, the *skin, flesh* (*C*, fig. 41), *core* (*D*, fig. 41), seeds (*E*) or *stones, stems* (*A*), and in kernel fruits the *calyx* (*B*), have all, in some cases, marked peculiarities, and in others more minute and scarcely perceptible; but yet in a strictly scientific study of pomology, of more or less service. It would be foreign to the purposes of this work to notice these points in detail; all that is deemed necessary, useful, or appropriate, is to point out well-defined and practical distinctions, and the terms ordinarily made use of in popular descriptions.

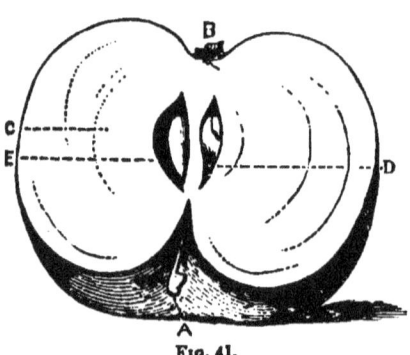

Fig. 41.

Fig. 41, vertical section of an apple, showing its different parts. *A*, the base. *B*, the eye. *C*, the flesh. *D*, the core. *E*, the seed. *A*, stem. *B*, calyx.

3d. *Different Parts of the Fruit:*

The Base (*A*) is the end in which the stem is inserted.

The Eye (*B*) is the opposite end, in the apple, pear, etc., that have an adhering calyx.

The Neck, in pears, the contracted part near the stalk, as seen in fig. 49.

The Point is the end opposite the stem in stone fruits;

berries, etc., that have no calyx, and consequently no eye.

The Length is the distance from stem to point or eye, A to B, fig. 41.

The Width, the line D E—cutting the fruit across, or at right angles with the length.

The Basin, the depression around the eye or calyx in kernel fruit, B, fig. 41.

The Cavity, the depression around the stem.

The Suture, in stone fruits the furrow-like depression running from the base to the point.

4th. *Different Properties of Fruits*:

Besides the principal divisions which have been alluded to, fruits are considered in regard to their *size, color, form, texture, flavor*, and *season of ripening*.

1st. *The Size*.—Besides the natural difference in size that exists among different varieties of the same species, as, for instance, between the *Bartlett* and *Seckel* Pears, or the *Fall Pippin* and *Lady Apples*, there are great differences between the same varieties owing chiefly to the following circumstances: *Soil*—We find that in new and fresh soils, the nutritive properties of which have not been impaired by cultivation, as in the virgin soils of the West, fruit of the same variety attains nearly *double the size* that it does in older parts of the country, where the soil has long been under cultivation; and that in the same orchard, the tree growing in a deep, alluvial soil, will give fruit much larger than the one on a hard gravelly knoll. *Culture*—This has an important influence on the size of fruits. If an orchard has been for several years neglected, and the ground about the trees become covered with grass and weeds, the fruit is small; and if the same orchard be ploughed up, some manure turned in around the roots, and the ground be kept loose and clean by tillage, the fruit will double in size in a single season. *Seasons*—In a dry season, when the supply of moisture at the roots and in the atmosphere is very limited, fruits are invariably smaller than in seasons of an opposite character. *Number of fruits on the tree*—This affects the size of the fruit to a great extent in all seasons, soils, and climates, and under all grades of culture.

It is perfectly obvious, that the greater the number of fruits a tree bears, the smaller they will be, for as they derive their sustenance from the tree, a large number cannot be so well supplied as a smaller number. We cannot go into an orchard where there are many varieties without seeing an illustration of this. Here is a prolific variety *loaded* in every part; the fruits are small, certainly not over medium size. There is a moderate bearer; its fruits are thinly and evenly distributed over the tree; its fruits are consequently *large*. So in the case of fruits that have been thinned; that is, a certain portion removed while young, either by accidental circumstances or by design, every specimen is twice as large, as if the whole crop had been allowed to mature. The English gooseberry *growers*, in preparing their prize specimens, leave but a few on each bush—not over a twentieth, or perhaps a fiftieth part of the entire crop. So in peaches, grapes, etc., grown carefully in houses. Where the size and beauty of the fruit, and the health and vigor of the trees are kept in view, a large portion of the crops, from one half to two thirds, is thinned out before maturity. *Age of the trees*—This influences the size of fruits to a great extent; we see fruit so large on young trees as to be entirely out of character: As trees grow older, the vigor decreases, and the number of fruits increase, and they are consequently diminished in size. *The kind of stock* has a tendency to modify the size; thus we find many pears much larger on the *Quince* stock than on the pear, and many apples larger on the *Paradise* than on the common apple stock. The reason of this is, no doubt, that on the quince and paradise the juices of the tree are better prepared, richer, and better suited to the growth of the fruit. In the common pear and apple stocks the sap is taken up in greater quantities, is watery, and better adapted to form wood than fruit.

CLASSIFICATION OF SIZE.

The terms qualifying the sizes of fruits are always given comparatively, in regard to the two extremes, the *largest* and the *smallest* of the species; for instance—in apples, we may consider the *Gloria Mundi* and *Twenty Ounce* as *extremely large*, and the Lady apple as *extremely small*. The terms used, therefore, are such as to represent the various grades between the two extremes. These are

Very large, as the *Gloria Mundi* Apple, *Duchesse d'Angoulême* Pear, *Crawford's Early* Peach, *Yellow Egg* Plum, and *Napoleon Bigarreau* Cherry.

Large, as the *Baldwin* Apple, *Bartlett* Pear, *Red Cheek Melocoton* Peach, *Washington* Plum, and *Black Eagle* Cherry.

Medium, as the *Rambo* Apple, *White Doyenne* Pear, *Imperial Gage* Plum, and the *American Amber* Cherry.

Small, as the *Early Strawberry* Apple, *Dearborn's Seedling* Pear, *Green Gage* Plum, and *Bauman's May* Cherry.

Very Small, as the *Amire Johannet* Pear, *Lady Apple*, *Winter Damson* Plum, and the *Indulle* (*Early May*) Cherry.

The distance between some of these grades, as between medium and large, &c., is so short that they are frequently confounded; still they give a notion of comparative size that answers all practical purposes. It would, perhaps, have been more accurate, and, at the same time, more satisfactory to persons entirely unacquainted with fruits to have given the comparative measurement of these different grades in inches and parts; but the varieties quoted as examples are common, and very generally known.

2d. *Form.*—It is exceedingly difficult, even impossible, to find any single term that will give a mathematically accurate notion of the forms of fruits; for although we call an apple round or conical, it may not be, strictly speaking, either; perhaps partakes to some extent of both forms. But that is no reason why we should designate it *conical round;* we simply call it *round*, or *roundish*, if nearer round than any other form; and if it inclines slightly to the conical, we cannot so well convey the knowledge of that fact any other way as by simply saying so.

In the apple the *round* form prevails, and in the pear the pyramidal; hence, it is necessary to apply a different class of descriptive terms to each.

FORMS OF APPLES.

Round or Roundish (fig. 42).—When the outline is round, or nearly so, the length being about equal to the breadth.

Flat (fig. 45).—When the ends are compressed, and the width considerably greater than the length.

Conical (fig. 43).—In the form of a cone, tapering from the base to the eye.

Ovate, or *egg-shaped* (fig. 44).

Oblong (fig. 46).—When the length is considerably greater than the width, and the width about equal at both ends, not tapering as in the conical.

THE FRUIT. 41

In addition to these forms and their various modifications, some varieties are

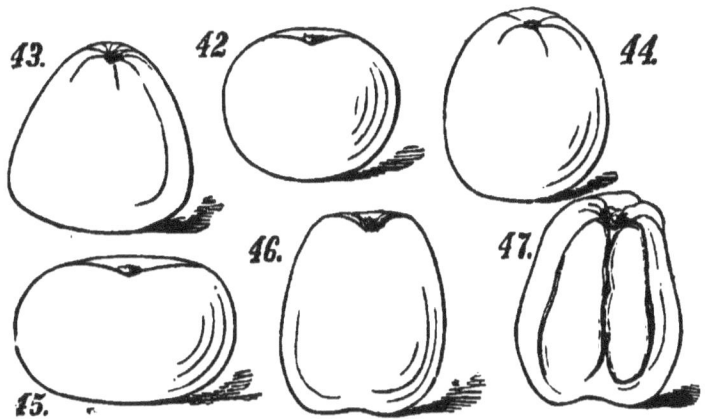

Figs 42 to 47, forms of apples. 42, round. 43, conical. 44, ovate. 45, flat. 46, oblong. 47, ribbed.

Angular, having projecting angles on the sides.
One-sided, having one side larger than the other.
Ribbed (47), when the surface presents a series of ridges and furrows running from eye to stem.

FORMS OF PEARS.

It has been remarked that the pyramidal form prevails in pears; but they taper from the eye to the stem, which is just the reverse of the tapering form in apples. Their forms are designated thus—

Pyriform.—When tapering from the eye to the base, and the sides more or less hollowed (concave) (fig. 48).

Long Pyriform.—When long and narrow, and tapering to a point at the stem (fig. 49).

Obtuse Pyriform.—When the small end is somewhat flattened (fig. 50).

Obovate or *egg-shaped.*—Nearly in form of an egg, the small end being nearest the stem (fig. 51).

Turbinate or *top-shaped*.—The sides somewhat rounded and tapering to a point at the stem (fig. 52).

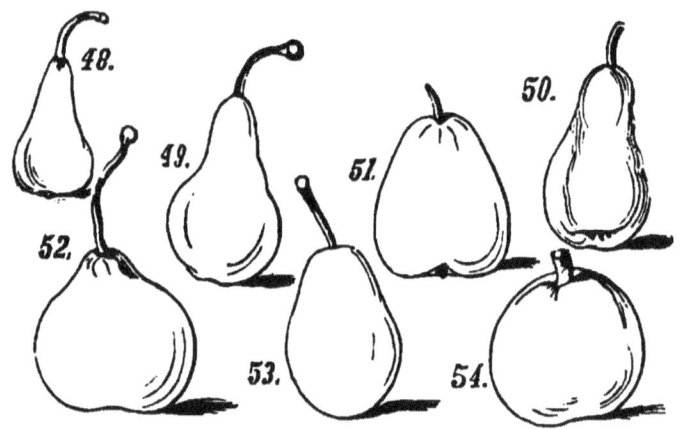

Figs. 48 to 54, forms of pears 48, pyriform. 49, long pyriform. 50, obtuse pyriform. 51, obovate. 52, turbinate. 53, oval. 54, round.

Oval.—Largest in the middle, tapering more or less to each end (fig. 53).

Round.—When the outline is nearly round (fig. 54).

FORMS OF PEACHES.

There is too much uniformity in the forms of peaches to render the adoption of any set of terms descriptive of them very serviceable. They are mostly round, occasionally approaching to *oblong* and *oval;* two sides are frequently compressed, flattened, exhibiting a suture or furrow running from the point to the base: the width, depth, &c., of this suture are, in many cases, peculiar, or at least worthy of note.

FORMS OF PLUMS.

Plums are *round*, *oval*, or *oblong*, as the peach, and marked, in some cases, by a similar flattening of the sides, and by the suture.

FORMS OF CHERRIES.

Cherries are *round* or *heart-shaped; obtuse heart-shaped*, when too round to be fully heart-shaped; and *pointed*, when the point is more than ordinarily sharp or peaked. The suture is also taken note of as in plums and peaches.

Gooseberries and *Grapes* are always round or oval. *Currants* always round. *Strawberries* round, conical, or oval, sometimes with a neck; that is, the base is drawn out at the stem in the form of a narrow neck. *Raspberries* are conical, roundish, or long.

3d. *Color.*—The color of fruits depends much on their exposure to the sun's rays. We find that in orchard trees, where the heads are dense, and a large portion of the fruit shaded and shut out from the sun, there is a great difference in the color; indeed, so great, frequently, as to make their identity from appearance quite doubtful. Varieties that are naturally—when properly exposed to the sun—of a bright red or a glowing crimson, remain green in the shade. The climate, too, seems to have considerable effect on the color. As a general thing, we observe that northern apples are clearer and brighter colored than those of the south.

Dry soils and elevated situations produce more highly-colored fruit than damp and low valleys. The terms used in describing colors, are all simple and well understood.

4th. *Flavor*, in table fruits, is one of the most important of qualities; for however large or fair a fruit may be, if insipid or astringent to an unpleasant degree, or if it possesses some other disagreeable quality, it is unfit for the table. There are various kinds of flavor even among varieties of the same species: in pears, particularly, it is almost endless, the shades and degrees of *sweet* and *acid*, and the various perfumes that mingle with these, are almost infinite.

The same circumstances mentioned as favorable to high and brilliant coloring, are also favorable to the production of fine flavor. *Light, heat,* a *dry soil,* and *moderate growth,* seem to be all essential to fine flavor. On trees somewhat advanced in age, fruits are apt to be higher flavored than on young trees that have just commenced bearing, and in a dry than a wet season. The philosophy of all this is, that in a damp soil or season, or in a shaded situation, when trees are young and growing rapidly, the fruit receives more sap from the tree than can be properly elaborated by the action of the sun and atmosphere on its surface, and, consequently, the sugary principle is produced in small quantities—the juice is watery, sour, or insipid, as the case may be.

The various terms by which flavor is designated, such as *sweet, acid, sub-acid, sprightly, perfumed, musky, spicy,* &c., are all well understood.

Section 8.—The Seed.

The perfect seed contains the rudiments of a plant of the same nature as that which produced it. This rudiment of the new plant is called the *embryo.* It consists of three parts—the *cotyledons* (*c c,* fig. 55), which are the rudiments of the first pair of leaves; these are the parts that first make their appearance. The bases of these cotyledons are united, and send down the *radicle* (*b*), or *root,* and between them is a bud (*a*), which sends up the stem, and is usually called the *plumule.* As soon as the seed is excited into germination by the heat and moisture of the earth, this *radicle* or root begins to penetrate the soil, and the plumule ascends in an opposite direction; and thus the growth of the tree goes forward

THE SEED.

in the manner already described under the heading, Root, Stem, &c.

It has been remarked that seed contains the rudiment of a plant similar to that on which it is produced; but this needs some explanation. In distinct *species*, this will be true; but the seeds of *varieties* that have been produced by culture and hybridizing, seldom or never reproduce exactly their like, hence the necessity for the various artificial methods of multiplication, such as grafting, budding, layering, &c. It is to these operations that we are indebted for the preservation of varieties that were originated hundreds of years ago.

Fig. 55.

Fig. 55, germinating seed. *a*, plumule. *b*, radicle. *c c*, cotyledons.

Germination.—Heat and moisture, air, and the exclusion of light, are all necessary to the healthy and perfect germination of seeds. It may be well to consider, briefly, the part which each of these has to perform.

1st. *Moisture.*—When seeds are sown in a time when the ground is parched, they will show no signs of germination until it is, in some way or other, moistened. The quantity of moisture necessary to a seed depends on the nature of its covering and its size. A small seed, with a thin covering, will vegetate much sooner and with less moisture than a large seed, with a hard, bony covering. The moisture must, in the first place, soften the covering, penetrate to the mealy part of the seed, and prepare it for the chemical changes necessary to convert it into food for the embryo plant. If apple or pear seeds be kept in a dry, warm room all winter, they will not be likely to vegetate the following spring, but if sown will probably lie in the ground all summer, and possibly germinate the spring following. If cherry seeds are kept dry for any length of time, say two or three months, they will not germinate the season following; and peaches and plums have actually to be in the ground all winter, under the action of frost, to insure their germination the spring following. Seeds will germinate much quicker when freshly gathered than after they have dried, because heat, moisture, and air have easier access to them, and act more quickly on them. These facts, of which all are well aware, show the necessity for moisture and the nature of its influence.

2d. *Heat* is the next most important element. Seeds do not grow in winter. We sow our apple, pear, peach, and plum seeds in November; but they show no signs of germination until a change of season. When the warmth of spring penetrates the soil, it reaches the seed, and, in connection with the moisture already imbibed, induces what we usually call fermentation. This chemical process excites the vital energies of the germ, decomposes the mealy part of the seed, and prepares it for the temporary nutrition of the young plant.

3d. *Air.*—Although seeds may have heat and moisture in the requisite proportions, still it has been proved by many experiments, that *without air*, germination cannot take place.

Practical cultivators are aware that seeds planted too deeply do not grow; many kinds will lie buried in the ground for years without growing, and when turned up near the surface will germinate immediately. It is the *oxygen* of the air that constitutes its importance; it produces, by forming new combinations with the gases contained in the seed, that chemical process which converts the starch into sugar and gum, as we observe in ordinary cases of fermentation.

4th. *Exclusion of Light.*—The manner in which self-sown seeds in the forest are covered with fallen and decaying foliage, plainly indicates that nature never intended the light to strike germinating seeds. A seed entirely exposed would be at one time saturated with moisture, and at another parched with drought; chemical changes would be alternately promoted and checked, until the vital principle would be destroyed, or so weakened as to produce a feeble and worthless plant. The depth of the covering should always be regulated by the size of the seeds. Small and delicate seeds may be sown almost on the surface, whilst large ones may be imbedded to the depth of four or five inches. The small seed requires little moisture, and has but a feeble force to penetrate an earthy covering; but the large requires much moisture, and has force enough to push its way up.

CHAPTER II.

SOILS.

SECTION 1.—DIFFERENT KINDS OF SOIL.

SOILS are usually designated by terms expressive of the predominant material in their composition, thus we hear of *sandy, loamy, gravelly, clayey, calcareous* or *chalky,* and *alluvial* soils.

A sandy soil is that in which sand is the principal ingredient. Such soil is usually quite defective. It is so porous that it parts almost instantaneously with moisture, and plants in it suffer from drought. All the soluble parts of manures are also quickly washed out of it, and hence it requires continual additions to produce even a scanty growth. The great point in improving it, is to render it more retentive by the addition of clay, ashes, &c.

A clayey soil is that in which clay predominates. It may be considered the opposite of sandy, inasmuch as its defects are, that it retains moisture too long, is too adhesive, in dry weather it becomes as hard as a burnt brick, impervious to dews or light showers, and when thoroughly saturated with wet it is tough, and requires a long time to dry. No fruit tree succeeds well in such a soil; but it is capable of being improved and fitted for many species, and especially the plum and the pear. The obvious way to improve it is, by incorporating with it lighter porous soils, as sand, muck, or leaf mould.

A gravelly soil is one made up in greater part of small stones, pebbles, decomposed rock, &c.; such soils, as a general thing, are unfit for fruit trees, unless great labor is incurred in trenching, deepening, and mixing with clay, muck, &c., of opposite characters.

A loamy soil is one we hear a great deal about, and may be understood in various ways. It may be considered a mixture of equal parts of sandy, clayey, and vegetable soil. It is neither so light as the sandy, on the one hand, nor so tenacious as clay on the other; and, as a general thing, contains such elements, and is of such a texture, as to render it eligible for all ordinary purposes of cultivation, and especially so for fruit trees. Loamy soils are spoken of as *sandy loams*, when sand forms a large ingredient, say one half of their composition; *gravelly*, when pretty largely mixed with small stones; *calcareous*, when lime is found in them.

Calcareous or chalky soils have a large amount of lime mixed with the other ingredients of which they are composed. All the lands in limestone districts are of this character, and, as a general thing, are well adapted to fruit culture.

Peaty soil consists chiefly of vegetable mould from decayed marsh plants, in low, wet places. It is unfit, in itself, for fruit trees, but is valuable for improving both light and heavy soils.

Alluvial soils are composed of decomposed vegetable substances, the sediment of rivers, and materials washed down from neighboring hills; the valleys of all our rivers and streams are composed of this, and it is the richest of all soils. Fruit trees in such soils make a rank, vigorous growth, but they are not so hardy nor so fruitful, nor is the fruit so high flavored as on soils with more sand, clay, or gravel, and less vegetable mould.

In treating of the different classes of fruits, we shall refer to the particular soils best adapted to them.

SECTION 2.—DIFFERENT MODES OF IMPROVING SOILS.

In regard to depth, soils vary materially, some being not over eight or ten inches in depth of surface, others a foot, while in deep alluvial valleys they are often two feet. For orchard and garden purposes, a *deep soil* is quite essential to enable the roots to penetrate freely in search of food, and to enable them to withstand the demands of protracted droughts. Few soils in their ordinary condition of farm culture are, in this respect, suitable for trees. Even where naturally deep and loamy, if the upper part only (say to the depth of six inches, which is as deep as most people plough) be in a friable condition, it cannot be considered as in a proper state for the reception of trees, for their roots cannot be confined to six inches of the surface. Some means of loosening and deepening must be resorted to, and what are they?

1. SUBSOIL PLOUGHING.

This is the cheapest and best method, where a large quantity of ground is to be prepared for extensive planting. The common plough goes first, and takes as deep a furrow, as practicable. The subsoiler follows in the same furrow and loosens, without turning up, the lower part of the surface and a part of the subsoil. Except in cases where the subsoil is a very stiff clay or a hard gravel and near the surface, the two ploughs can go to the depth of eighteen or twenty inches. This is our mode of preparing nursery grounds. If a single ploughing in this way does not accomplish the desired end, a second may be given, going down still deeper.

SOILS. 51

We had a piece of soil the surface of which was about a foot deep of black vegetable mould, with a slight admixture of sand, resting on a stiff clay subsoil, which prevented the water from passing off. In this condition we found it entirely unfit for trees; we subsoil ploughed it six or eight inches deep, turning up the clay subsoil and mixing it with the surface; we also drained it, and spread over the surface the clay that came out of the drains, and in this condition we find it producing the finest trees, especially apples, pears, and plums. The soil is more substantial, and the surface water passes off freely.

2. TRENCHING.

In gardens too limited in extent to admit of ploughs, or where it is desired to make the soil thorough and permanently deep, trenching is the means.

The spade is the implement used in this operation. A trench two feet wide is opened on one side of the ground, and the earth taken out of it is carried to the opposite side. Another trench is opened, the surface spadeful being thrown in the bottom, and the next lower on the top of that, and so on till it is opened the required depth, which, for a good fruit garden, should be about two feet. If the subsoil be poor and gravelly, it is better to loosen it up thoroughly with a pick, and let it remain, than to throw it out on the surface. When the whole plot is trenched over in this way, the earth taken out of the first trench will fill up the last one, and the work is done. If the soil be poor, a layer of well-decomposed manure may be added alternately with the layers of earth; and if the soil be too light and sandy, clay, ashes, etc., can be added; and if too heavy, sand, lime, muck, peat, scrapings of dead leaves from the woods, or any other material calculated to render it porous and friable. If a garden is thus trenched in the fall or winter, and then turned over once in the spring to effect a thorough mixture of all the materials, it will be in suitable order for planting. This is something like the way to prepare soil for a garden; and let no one say it is too troublesome or too expensive, for in two years the extra pleasure and profit it will yield, will pay for all. Nothing is so expensive or so troublesome as an ill-prepared soil.

3. DRAINING.

There is a false notion very prevalent among people, that where water does not lodge on the surface of a soil, it is "dry enough." However this may be in regard to meadows or annual crops, it is quite erroneous when applied to orchards or fruit gardens. *Stagnant moisture* either in the surface or subsoil is highly injurious—ruinous to fruit trees. In such situations we invariably find them unthrifty and unfruitful, the bark mossy, and the fruit imperfect and insipid. All the soils, then, not perfectly free from stagnant moisture, both above and below, should be *drained*. In draining, it is, of course, necessary to have a fall or outlet for the water. Having selected this, the next point is to open the drains. We usually make them three feet deep, and wide enough to give sufficient room to work—say three feet wide at top, narrowing gradually to six inches at the bottom, which should be even and sloping enough to the outlet of the water to enable it to run. A laborer who understands draining, will make two rods of these in a day; and good pipe-tile, two inches wide, can be had at the rate of about one cent per foot. Draining, therefore, is not so costly an operation as many suppose.

Where draining tiles are not to be had conveniently, small stones may be used. The bottom of the drain should be filled with them to the depth of eight or ten inches. In using these, the drains require to be at least six inches deeper than for tiles, in order that a sufficient quantity of stones can be used without coming too near the surface. Some brush, or turf, with the grassy side downwards, should be laid on the stones before filling in the earth, to keep it from filling up the crevices.

CHAPTER III.

MANURES.

Section 1.—Importance of Manures.

No soil, whatever may be its original fertility, can sustain a heavy and continued vegetation for many years without becoming, to some extent, exhausted. Indeed, there are few people so fortunate, except those who settle upon new, uncultivated lands, as to procure a soil that does not need manuring to fit it for the first planting with trees. It is, then, a matter of importance for every man who has more or less land to cultivate, to inform himself well on the subject of saving, preparing, and applying manures. In this country, the only class of men, generally speaking, who can be properly said to collect and manage manures with system and care, are nurserymen and market gardeners near our large towns. It is very seldom that people generally give the matter a thought until garden-making time comes around in the spring; and then, anything in the form of manure is carried into the garden, and applied whether fit or unfit. This is not the proper course.

Every garden should have its manure heap, that, in the fall or spring, when it comes to be applied, will cut like *paste*. In that state only is it safe to apply it. All parts of it are then decomposed thoroughly; all seeds of noxious plants are dead, and it is in a condition capable of yielding at once, to the roots of growing plants, nealthy nutrition, that will produce a *vigorous, firm, sound*, and *fruitful growth;* and this is precisely what is wanted: far better to have a tree starved and stunted, than forced into a rank, plethoric growth, with crude, ill-prepared manures.

SECTION 2.—PREPARATION OF MANURES.

The best gardeners pursue a system something like this: A trench is prepared two or three feet deep, and large enough to hold what manure may be wanted. In the bottom of this trench, a layer of muck, grassy turf, ashes, anything and everything capable of being decomposed, is laid down, say a foot deep. On the top of this, a thick layer of stable or barnyard manure, two or three feet deep, then another layer of muck, gypsum, etc. In this way it remains till more manure has accumulated around the stables; it is then carried and deposited in another layer, with a layer of the other materials on the top. The manure should always be saturated with moisture, and trodden down firmly to hasten its decay, and if an occasional load of night soil could be mixed in with it all the better. The layer of muck and other substances being always placed on the top of the last layer of manure absorbs the evaporations of the heap, and hastens the decay of all. When stable manure is thrown down and left uncovered, a dense steam will be seen to rise from it; and this is the very essence of it escaping to be lost, and if it be thrown down in a heap *dry* it will immediately burn—that is, dry rot. Its enriching ingredients all pass off by evaporation, and there is nothing left but its ashes, so to speak.

When the heap has accumulated for four or five months as described, the whole should be turned over, completely mixed, and piled up in a compact, firmly-trodden mass, when it will undergo farther decomposition and, in a short time, become like *paste*. Adjoining every manure heap, there should be an excavation to receive its liquid drainage, in order that it may be saved, and either applied in the growing season, in a liquid state, diluted with water, or be thrown over the heap.

"Special manures" have been much talked of lately. By the word "special," is meant a particular quantity, of a particular mixture, for certain species, and even for certain varieties of fruits. Nearly all the suggestions on the subject are speculative and unreliable. The subject is an important one, but we want direct and careful experiments. It is only when we know to a certainty what material certain trees need most of, and in what degree it abounds or is wanting in our soil, that we can apply it safely. The experience of farmers and gardeners, grain and fruit growers, all over the world, affords undoubted evidence of the enriching qualities of *stable manure*. On all soils, and for all sorts of crops, it is an unfailing and powerful fertilizer; and we make it the base of all our manure and compost heaps. By mixing with it the ingredients we have mentioned, we hasten its decay, save its parts from waste, and, at the same time, combine with it other substances that will not only enrich but improve the texture of soils, and increase the supply of the mineral substances required by plants. Dr. Daubney, a distinguished writer on the character and improvement of soils, etc., says, "Fortunately we are provided in the dung of animals with a species of manure of which the land can never be said to tire, for this simple reason, that it contains within itself not *one* alone, but *all* the ingredients which plants require for their nutrition, and that, too, existing in the precise condition in which they are most readily taken in and assimilated." But a good substitute for this article, where it cannot be obtained, is an important point. Some time ago, we noticed in the report of a discussion on manures in Boston, that the Hon. M. P. Wilder, one of the most distinguished horticulturists in America, stated that he had found the following compost equal to stable manure for gardening purposes generally, and for fruit trees.

"One cord of meadow muck, having been exposed to the action of the air and frost at least one year; twelve bushels leached ashes; six bushels crushed bones. This mixture cost him at the rate of $4 50 cents per cord. Latterly he added to this his stable manure, and about an eighth of the whole bulk of fine refuse charcoal from the depôt of venders, which was delivered to him at $5 per cord; and in this way he found it the best, as a general manure, he had ever used. On fruit trees its effect was remarkable.

"In the spring of 1847, he planted a square in the nursery with imported trees from England, this compost having been spread and ploughed in. These trees were from four to five feet in height, and although it is not usual for trees to make a large growth the first year, they acquired branches of three to four feet, and were so handsome as to command $1 25 each, for a row of fifty trees, without any selection.

"In June last, which is very late to set out trees, he prepared another square on rather poor land, and planted trees just received from England upon it. The soil had been thrown up to the frost the previous winter, and the compost here was applied in the trenches near the roots. Mr. Wilder exhibited two shoots which had grown from those trees since they were set in June. The shoots were four feet in length, and the wood hard and well ripened."

In addition to all these sources for manure, it may be added that fallen leaves, scrapings of streets, weeds, wood chips, sawdust, the ashes of all prunings of trees and brush, soot, blood, animal flesh, soap suds, and slops from the kitchen, and, in fact, everything decomposable may be used, to increase the bulk of the manure heap, taking care that everything likely to waste by evaporation be covered at once with muck, charcoal, or some material calculated to absorb the gases evolved by decomposition. We very frequently see people, in the spring of the year, when their garden is undergoing a purifying and fitting up process, carry to the highway all the brush, dry stems of plants, and all the wreck of the previous season's work, there to make a bonfire to get it out of the way, while at the same moment they complain sadly of the lack of manure.

There was no such thing as a manure heap on the premises.

SECTION 3.—MODES OF APPLYING MANURE.

Where an acre or several acres of ground are to be prepared for trees, the better way is to spread the manure over the surface and turn it in with the plough. When it is scarce and economy necessary, it may be applied around the roots, by mixing with the earth at planting time.

Quantity to be Applied.—This, of course, depends on two things, the necessities of the soil and the quality of the manure. If the land be poor, an even covering of two or three inches should be given; if in tolerable good condition, one inch will be sufficient. One inch of well decomposed animal manure will be equal to three inches of a partially decayed compost.

SECTION 4.—LIQUID MANURE.

Manure in a liquid state has these advantages to recommend it. It can be applied to trees and plants in a growing state without in the least disturbing the surface of the soil, and it supplies, at the same time, both nutriment and moisture. It can be applied to bearing trees, strawberries, etc., *in fruit*, if defective in vigor, or suffering from drought, and yield an immediate sustenance that will enable them to produce much larger and finer fruit than they could have done without it.

It may either be collected in a tank, kept on purpose near the barns, or it may be made when wanted by dissolving manure in water. It may be much stronger for trees, the roots of which are a considerable distance from the surface, than for such plants as have their roots near the surface. It is the only prompt and effectual stimulant for trees on a poor soil, to enable them to perfect their crop. We have frequently witnessed its astonishing effects. It should be applied in the evening, and in such quantity as to penetrate to the roots; half a dozen waterings will be sufficient in most cases, but it is better to apply it well diluted and often, than a smaller quantity too strong. A dozen shovelfuls of animal manure will make a barrel of liquid powerful enough for most purposes; and if pure liquid soakage of the manure heap or urine of animals is used, at least one half rain water should be added. Soap suds form an excellent liquid manure for all trees. The grape vine is especially benefited by liberal and frequent application.

CHAPTER IV.

THE DIFFERENT MODES OF PROPAGATING FRUIT TREES.

General Remarks.—The propagation of fruit trees may be classed under two principal heads—the *Natural*, which is by *seeds;* and the *Artificial*, by the division of the *plants*, as in *cuttings, layers, suckers, buds, and grafts.*

PROPAGATION BY SEEDS.

Seedling fruit trees are propagated, either to obtain new varieties, or stocks for budding or grafting. It is only where the very rudest system of fruit culture is practised, as for instance in newly-settled countries, that seedlings are planted out to bear, for the reason that, unless in very rare instances, varieties worthy of cultivation do not reproduce themselves from seed. The important differences that exist between the seeds of different classes of fruit trees, render it necessary to treat of each separately; their management will therefore be given in detail, in connection with the propagation of stocks.

There are some points, however, of general application that may be considered here with propriety. It scarcely admits of a doubt, but that the greater part of the difficulties met with in fruit tree culture, as maladies of various sorts, unfruitfulness, etc., are induced by a careless and undiscriminating system of propagation. The stock has a most important influence on the health, longevity, fruitfulness, and symmetry of trees, and it does not seem possible that our indiscriminate mode of saving seeds for stocks is at all consistent with rational, intelligent culture.

What is the ordinary course? To raise apple seedlings, a quantity of pomace is procured at the cider mill, without the least regard to the quality or maturity of the fruits from which it was produced, or of the health, vigor, and hardiness of the trees that bore the fruit—these points are never thought of. So it is in the case of pear seeds. During the last few years, these seeds have been nearly as valuable as gold dust; the price being seldom less than $5 per quart. The present season, a neighboring nurseryman has paid at the rate of $4 per quart for a bushel. How is this seed procured? Is it *selected* from healthy, vigorous trees, with sound constitutions, and from perfect, well-matured fruits? By chance it may be; but seed collectors are usually glad to find fruits of *any kind*, and from any *sort of tree*, if they have only seeds apparently good. We do not, by any means, intend to charge upon any man a fraudulent intent in this matter. The seed collector is no more to blame than the nurseryman, for the nurseryman seldom asks any particulars about the origin of the seeds. How is it with peach trees? The peach is a short-lived tree, highly susceptible of deterioration from bad treatment; and it is obviously impossible for an unhealthy, feeble tree, to produce sound and healthy plants from their seeds. In some districts of the country, a sound, vigorous peach tree is a rarity; and yet, how are peach seeds saved and procured? The seeds are brought in to the seedsman, he buys them without asking any questions about either the health or sickness of the trees that produced them. They are peach stones, and that is all necessary to be known. The nurseryman buys of the seeds man just as he received them; this is the way that the country has been filled with miserable, diseased, and unsightly trees, and who is in the fault? "Why," most people would say, the "*nurserymen,* of course. They ought to be more careful in selecting their seeds, so that they might be certain of having sound and healthy stocks. They ought to select the fruits, from which to obtain their seeds, while on the tree, and see that the trees are not in an incipient, or, perhaps, an advanced state of decay, but in full health and vigor, possessing such characters, as to habit, growth, and hardiness, as are desirable in the best quality of nursery stock." Very true, it must be admitted. This is precisely the course that nurserymen ought to pursue. It is the course followed in the great orchard districts of France, and that ought to be adopted everywhere. But we must have *cheap, easy,* and *labor-saving* modes of doing things now-a-days; as well the raising of trees as everything else. Suppose a nurseryman could be found who would go about the culture of trees after some such system as we have indicated, it must be very clear that he could not sell his trees as cheap as another, who followed the present almost universal hap-hazard course, and if he could not do this, the probability is he would be compelled to keep them; for purchasers of trees, as a general thing, make no such discriminations. It happened one season that more than the usual quantities of seedling, unworked, peach trees were brought into the streets of Rochester for sale; they were as miserable, in *all respects*, as trees could be; yet they were sold by the thousand, at from 4 to 8 cents apiece, and scarcely one of them ever grew, for they were killed by exposure, *fortunately.* At that very time there were large stocks in the nurseries, about town, of good worked trees of the best varieties, offered at *one shilling each.* This instance is quoted simply to show who are to blame for the defective and vicious systems of propagation usually practised. That there will be a reform soon is not to be doubted. A discriminating spirit is already becoming apparent among the best classes of cultivators, and their example will soon be felt. The selection of seeds for stocks is a point of more than ordinary importance, and merits the special attention of every man engaged, to whatever extent, in the propagation of fruit trees.

Production of New Varieties.—New varieties are produced from seeds that have been properly hybridized, as described in the article on hybridization, or from seeds of the best specimens of the best varieties.

Where it is desired to obtain seedlings of a particular variety, free from any crossing with others, the flowers should be protected while in blossom, to guard them against foreign impregnation; seeds should be saved only from large, perfect, fine flavored specimens, and the seeds themselves should be plump and mature. Sometimes a good variety is obtained by selecting from beds of seedlings, such as possess marked evidences of improvement, *vigorous, luxuriant growth, large heavy foliage, prominent buds,* and *smooth, thornless wood.* These characters indicate superiority, but do not always ensure superior fruit. The stock is supposed to exert considerable influence on the seed; and if this be the case, it would be well to get such varieties as we wish seeds from, on their own roots, by layering, or grafting on roots in the ground, so that the graft will itself strike root. Mr. Knight's mode of obtaining seedlings, of the best varieties, was to prepare stocks from some good sort that would strike from cuttings. These stocks he planted in rich warm soil, and grafted with the kind he wanted the seeds from. The first season after grafting he took them up, reduced the roots, and planted again. In this way he had them bear fruit in two years. He allowed only a couple of specimens to remain on each tree, and these, consequently, were very large, mature, and every way fine, and from these the seeds were taken. Seedlings may be tested quickly, by budding or grafting them on bearing trees. We may fruit apples and pears in this way, in four or five years, whilst ten or fifteen would be necessary on their own roots. Experimenters on this subject have found the seeds of new varieties are more certain to produce good fruit than the seed of old ones.

2. *By Division of the Plants.*—It has been remarked in the article on buds, that every bud is capable, under favorable circumstances, of producing a new individual, similar to that from which it is taken.

Hence it is, that out of the young annual wood of an apple, pear, peach, or any other fruit tree, we frequently make several hundreds. Every good, well-formed bud, properly separated, and inserted under the bark of the individuals of the same, or a closely allied, species, will, in one year from its insertion, or with one season's growth, have become a new tree. It is by these means we are enabled to disseminate new varieties with such wonderful rapidity. If a young tree of a new variety will make half a dozen shoots the first season, each bearing half a dozen buds, we can, if we have stocks to bud on, be in possession of thirty trees of that variety in two years from the time we obtained one tree, and in another year we may have four times that number. The production of a tree from a *bud*, a *graft*, a *layer*, or a *cutting*, is but the same thing effected by different means. In all the cases, a part of the parent plant, with one or more buds attached, is separated from it. The *cutting*, sometimes composed of one bud or joint, and sometimes of several, we put directly in the ground, where it forms roots. The graft is a cutting in-

PROPAGATION BY CUTTINGS.   65

seed, not in the ground, but in the wood of another plant to which it unites. The bud inserted under the bark of another tree, and the one buried in the ground, differ only in this, that one draws its support directly from the soil, and the other indirectly, through the tree to which it unites.

SECTION 1.—PROPAGATION BY CUTTINGS.

A cutting is a shoot, or part of a shoot, generally of one season's growth. The length of the cutting varies from a single eye or joint, to a foot, according to the nature of the species, or the circumstances under which they are to be grown. The wood should be as stout and mature as possible, and should be cut close and smooth to a bud at both ends (fig. 56). In all cases, cuttings taken off closely to the old wood, with the base attached, as in fig. 57, are

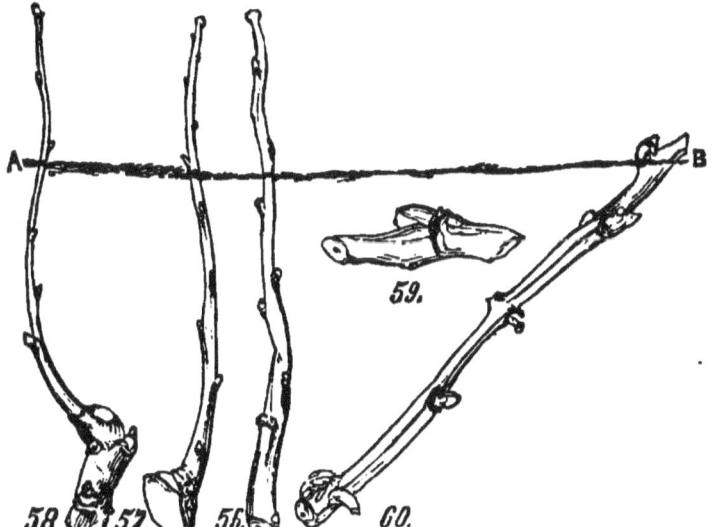

Figs. 56 to 59 cuttings. 56, a cutting, all of young wood. 57, a cutting, with a heel of old wood. 58, a cutting, with 2 or 3 eyes of old wood. 59, a cutting, of a single eye of the grape vine. 60, a long cutting of the grape, line *A, B*, surface of the ground.

more successful than when cut at several joints above; and in many cases, as in the quince for example, an inch or two of the old wood left attached to the base of the cutting, as in fig. 58, renders it still more certain of success The more buds we can get around the base of a cutting, the better, other things being equal; for these buds, as soon as they become active, send down new matter, from which the roots are emitted.

Cuttings of the grape are sometimes made of a single eye (fig. 59), with an inch or so of wood above and below it.

The *time to make cuttings* is in the fall, as soon as the wood is ripe, and through the early winter months. It should not be deferred later than January. The *soil for cuttings* is of the greatest importance to their success, for if, on the one hand, it be cold, damp, and compact, they will decay, and if too loose and sandy, they will dry up for the want of sufficient moisture. A soil so mellow that it cannot bake, and yet so compact as to retain humidity enough to support the cuttings, until new roots are formed, seems to be absolutely necessary—such a soil as we may suppose a good garden border to be composed of. Rooted plants can endure extremes, but cuttings require the most favorable circumstances.

Time to Plant.—The fall would be the better season to plant all cuttings, if we could cover them so as to prevent the frost from heaving them out. It is on account of this difficulty that we plant, from necessity, in the spring; but spring planting must be done very early, that vegetation may proceed gradually. If late planted, warm weather comes on them at once, before they have formed roots sufficient to support the demands of the young leaves. Where only a few are grown, shading might, at certain times, be given, and some light substance, like saw dust, be spread about them, to preserve an even temperature and humidity, or they might be put in a cold frame, where they could receive any required attention. Where acres of cuttings are grown, these things are not practicable.

Depth to plant.—As a general thing, cuttings should be inserted so deep, that only two buds will be above the surface of the ground, and in the vine only *one*. If cuttings are long, they need not be set perpendicular, but sloping, so as to be within reach of heat and air. A cutting of a single eye of the vine with a piece of wood attached, must be entirely covered, say half an inch deep; see figures 56 to 60, ground line, A, B. But such cuttings are seldom planted, except in pots, in houses, or in hotbeds.

Preserving Cuttings.—If cuttings are not planted in the autumn, they should at least be prepared quite early in the winter, and be buried in the earth out of doors, in a pit. A mound of earth should be drawn up over the pit to throw off water. At the very first favorable moment in the spring they should be planted. Trenches are opened as deep as necessary with a spade, and the cuttings set in it at the proper distance, from three inches to a foot, according to circumstances. When the cuttings are in the trench, the earth is partly filled in, and trod firmly down with the foot, then the balance is filled in and levelled up.

Cuttings require particular attention, in the way of weeding and hoeing; if weeds grow up thickly, and appropriate the moisture of the ground, or if the surface be allowed to crack, as it may after rains, if not quite sandy, they will either make a feeble growth, or fail entirely. The ground wants repeated stirring, to keep it friable and perfectly free from weeds.

Section 2.—Propagation by Layering.

A layer is similar to a cutting, except that it is allowed to remain in partial connection with the parent plant until it has emitted roots. On this account, layers are much more certain than cuttings. It is the best method of propagating the *grape* and the *gooseberry*, and also the *quince, paradise,* and *Doucain,* for stocks. It may be performed in the spring with shoots of the previous year's growth, before vegetation has commenced, or in July and August on wood of the same season's growth. The ordinary mode of doing it is, first, to spade over and prepare the ground in which the branch is to be laid, in order to make it light and friable. The branch is then brought down to the ground (fig. 61), an incision is made at the base of bud *A*, through the bark, and partly through the wood; the knife is drawn upward, splitting the shoot an inch or two in length, and the branch is laid in the earth with the cut open, and kept down by means of a crooked or hooked wooden peg, *B*. The earth is then drawn in smoothly around, covering it two or three inches deep; and the end of the shoot that is above ground, is tied up to a stick (*C*), if it requires support. In the *grape, gooseberry,* or *currant,* a simple notch below a bud is sufficient,

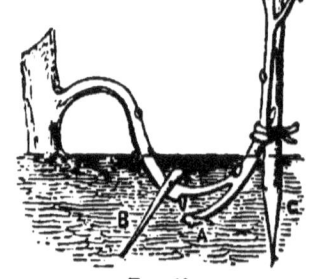

Fig. 61.

Fig. 61, a common layer. *A*, the incision. *B*, hooked peg. *C*, stake.

and they will root if simply pegged down; but roots are formed more rapidly when the shoot is cut one third through, and slit as described.

A long shoot of the vine may be layered at several points, and thus produce several rooted plants in the course of one season. This is called *serpentine layering* (fig. 62). The *Quince, Paradise*, and *Doucain* stocks, where raised in large quantities, are propagated in a different way from that described. The process requires much less labor; and where plants root so freely as they do, it answers every purpose.

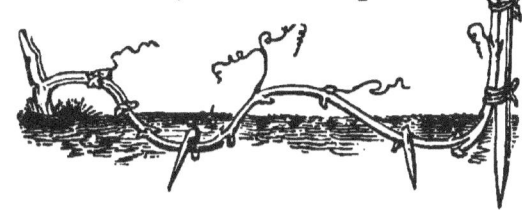

Fig. 62, Serpentine layer.

We will take a plant of the quince, for example, and, in the spring, before growth commences, we cut it down nearly to the ground, leaving four or five buds at its base (*A*, fig. 63). During that season, a number of vigorous shoots will be made. The following autumn or spring the earth is drawn up around the base of the plant, so that the crown where it was cut will be covered, and, consequently, the base of all the shoots for several inches in height. During the next summer's growth every branch is sufficiently rooted to be separated and placed in nursery rows the following spring. This is the way to obtain *strong* stocks; for the cutting back of the mother plant produces very vigorous shoots the first

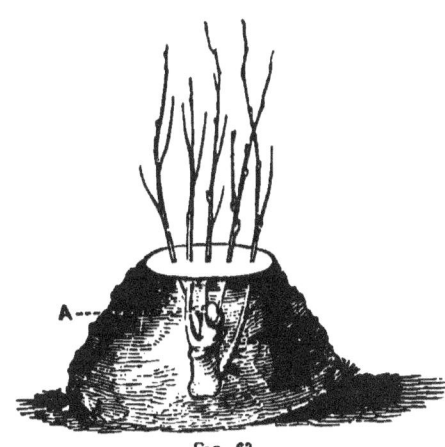

Fig. 63, Mound layering or banking up. *A*, the point at which the mother plant was cut back.

season, and when another season's growth is added they are as strong as can be desired. We succeed in rooting these shoots the first season of their growth by earthing them up about midsummer; but they are not quite strong enough, or sufficiently rooted, for transplanting and budding the following season.

Section 3.—Propagation by Suckers.

Suckers are shoots sent up from the roots. We observe them most frequently around trees that have had their roots wounded by the spade or plough. The wounds induce the formation of buds, and these buds send up shoots. They are occasionally used from necessity for stocks, but should not be employed where seedlings can be obtained. Occasionally we find certain varieties of plum throw up fine vigorous suckers, that would make excellent stocks if taken off with good roots; but their tendency to produce suckers renders them exceedingly annoying in gardens, and on this account objectionable. The roots of the raspberry are full of buds, and, consequently, throw up great quantities of suckers; and the smallest cuttings of the roots will grow. Suckers of any plants that can easily be propagated by cuttings or layers, should never be used.

Section 4.—Propagation by Budding.

This operation is performed during the growing season, and usually on young trees from one to five years old, with a smooth soft bark. It consists in separating a bud with a portion of bark attached, from a shoot of the current season's growth of one tree, and inserting it below the bark of another. When this bud begins to grow, all that part of the stock above it is cut away, the bud grows on, and eventually forms a tree of the same variety as that from which it was taken. Buds may be inserted in June, and make considerable growth the same season, but as a general thing this is not desirable in the propagation of fruit trees. The ordinary season in the Northern States is from the middle of July till the middle of September, and the earliness or lateness at which a species is budded depends, other things being equal, on the condition of its growth.

Those accomplishing their growth early in the season are budded early, and those that grow until the autumn are budded late—thus the season extends over a period exceeding two months. In all cases, the following conditions are necessary:

1st. *The buds must be perfectly developed in the axils of the leaves on the young shoots* intended to bud from. This is seldom the case until the shoot has temporarily ceased to lengthen, as indicated by the perfect formation of its terminal bud.

If buds are wanted before this condition naturally arrives, their maturity may be hastened very much by pinching the tips of the shoots. In ten or twelve days after the pinching of a very soft shoot, its buds are fit for working.

2d. *The bark must rise freely from the stocks to be budded.* This only happens when the stocks are in a thrifty and growing state. Where only a few stocks are to be worked, they can be easily watered, if necessary, a week or so before it is desirable to bud them. Trees that accomplish most of their growth early in the season, must be watched and budded before they cease to grow; those that grow very late, must not be budded early, or the formation of new wood will surround and cover the buds; in gardener's language, they will be "drowned by the sap."

The *implements* needed are a *pruning knife* to dress

the stocks, by removing any branches that may be in the way of inserting the bud; and a *budding knife* to take off the buds and make the incisions in the stock. The latter should have a very thin, smooth, and keen edge.

Strings for tying in the buds are either taken from bass mats, or they are prepared from the bark of the basswood. We always prepare our own; we send to the woods and strip the bark off the trees in June; we then put it in water from two to three weeks, according to the age of the bark, until its tissue is decomposed, and the fibrous, paper-like inner bark is easily separated from the outer, when it is torn into strips, dried, and put away for use. Before using, it should always be moistened to make it tough and pliable.

Cutting and Preparing the Buds.—Young shoots in the condition described, are cut below the lowest plump bud; an inch or two of the base of every shoot, where the buds are very close together, and quite small, should be left. The leaves are then stripped off, leaving half of each leaf stalk to handle the bud by, as in fig. 63.

Preserving the Buds.—When a considerable quantity is cut at once, they should be wrapped in a damp cloth as soon as cut and stripped of the leaves, and they may be preserved in good order for ten days, by keeping them in a cool cellar among damp saw-dust, or closely enveloped in damp cloths, matting, or moss. We often send buds a week's journey, packed in moss slightly moistened; the leaves being off, the evaporation is trifling, none in fact when packed up, consequently very little moisture is needed.

Having the stocks, buds, and implements in the condition described, the operation is performed in this way:

The shoot to bud from is taken in one hand, and the budding knife in the other, the lower part of the edge of the knife is placed on the shoot half an inch above the

PROPAGATION BY BUDDING.

bud to be removed (*A*, fig. 64), the thumb of the knife-hand rests on the shoot below the bud (*B*), a drawing cut is then made, parallel with the shoot, removing the bud and the bark to which it is attached, half an inch above, and three quarters below it. This is the usual length, but it may in many cases be shorter. The cut is made just deep enough to be below the bark, a small portion of the wood is always taken

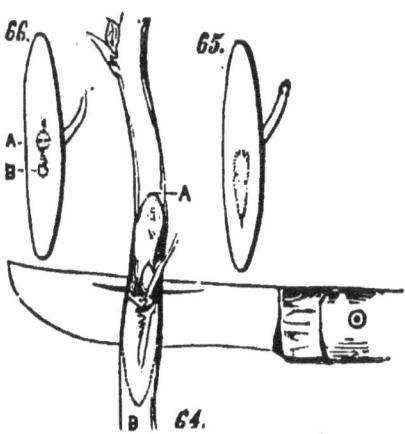

Figs. 64 to 70, Budding.

Fig. 64, a shoot of buds with the leaves taken off. *A*, the point above the bud where the knife was inserted. *B*, the point below where it comes out. *Fig.* 65, is a bud badly taken off, with a hollow in the centre. *Fig.* 66, a good bud. *A*, root of the bud. *B*, root of the leaf.

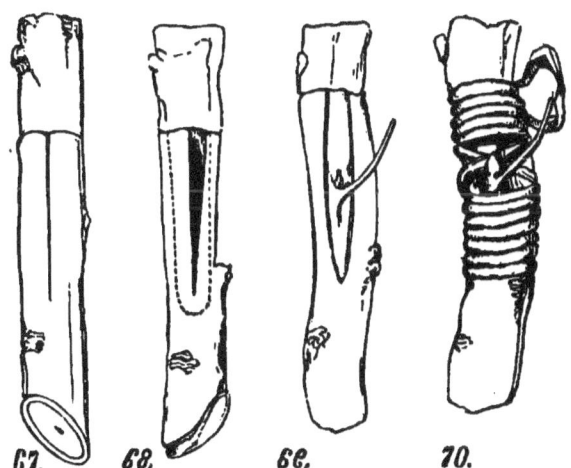

Fig. 67, a stock with the bark slit vertically and across. *Fig.* 68, the same with the bark raised as far as the dotted line. *Fig.* 69, the same with the bud inserted. *Fig.* 70, the same tied up.

off with it, and if this adheres firmly it should be allowed to remain; if it parts freely, it should be taken out, but in doing so the *root* of the bud must be carefully preserved, for if it comes out with the wood, the bud is useless. The root of the bud, as it is termed, is a small portion of wood in the hollow part of the inside of the bud. Fig. 64 is a good bud, *A*, root of bud, *B*, root of leaf. Fig. 65 is imperfect, the roots of leaf and bud both out. A smooth place on the stock, clear of branches, is then chosen. where two incisions are made to the depth of the bark, one across the end of the other, so as to form a T, fig. 67; the bark on the two edges of the perpendicular cut is raised (fig. 68) with the smooth ivory handle of the budding knife, and the bud is inserted between them (fig. 69); the upper end of the bark attached to the bud is cut square, to fit to the horizontal cut on the stock, the bass string is then wound around tightly, commencing at the bottom, and covering every part of the incision, leaving the bud itself, and the leaf-stalk, uncovered (fig. 70), the string is fastened above the horizontal cut, and the work is done. The success of the operation, as far as its execution is concerned, depends, in a great measure, on *smooth cuts*, an *exact fit of the bud to the incision made for it, secure, close tying*, that will completely exclude air and rain water, and the quick performance of the whole. The insertion of a bud should not, in any case, occupy more than *a minute;* ordinary practised budders will set two in that time, and often two hundred in an hour with a person to tie. Where the stocks and buds work well, two thousand is not an uncommon day's work in our nurseries, especially of cherries, peaches, and apples.

Where only a few buds are to be set, a cool, moist day or evening should be selected, as they will be more certain of success than if inserted during the middle of a hot, dry day.

The chief difficulty experienced by beginners in bud

ding, is the proper removal of the bud. When it happens that the knife passes exactly between the bark and wood, the bud cannot fail to be good; but this rarely happens—more or less wood is attached, and the removal of this is the nice point. Where the buds are flat, the difficulty is less than when they have large prominent shoulders, as the plum and pear have, in many cases. When all the wood is taken out of these, a cavity remains, which does not come in contact with the wood on which the bud is placed, and therefore, although the bark unites well, the bud will not grow. Sometimes, such as these are separated by making an incision through the bark; lift the edge of the bark attached to the bud with the knife, and push it off with the fingers. A safer way still is to cut around the bud, and draw a strong silk thread between the bark and wood, thus removing the bud in perfection.

SECTION 5.—PROPAGATION BY GRAFTING.

Grafting is the insertion of a scion of one species or variety on the stem or branch of another, which is called the *stock*. Its principal object is to increase certain varieties that cannot be reproduced from seed with certainty; but it is frequently performed with other objects in view. For instance—

To Fruit a New Variety.—A scion inserted in a branch of a bearing tree, will bear fruit perhaps the second year from the graft; but if the same scion had been put on a young seedling, it would not have borne in ten years.

One *species* is frequently grafted with success upon another, by which certain important modifications are wrought upon both the size and fruitfulness of trees, and the quality of the fruits. Thus, we can graft, in many cases, with highly beneficial results, the peach and apricot on the plum; the pear on the quince; strong growing species and varieties on weaker ones, and *vice versâ*. But experience has established the fact, that there must be between the stock and graft a close alliance. We cannot graft an *apple* on a *peach*, nor a *cherry* on a *pear;* but the pear, the apple, quince, medlar, thorn, and mountain ash—a naturally allied group—may, with more or less success, be worked upon one another.

The French horticulturists, who are the most skilful and curious in all matters pertaining to the propagation of plants, describe in their works upwards of one hundred different modes of grafting, practised in different ages and countries, and for the attainment of particular objects; but, however interesting the study of all these may be to the student and experimentalist, the great bulk of them are of little practical utility, and are never applied in the multiplication of fruit trees. It is, therefore, unnecessary to fill up the pages of such a treatise as this, with either a historical account or description of them. The methods described below are those universally adopted, with slight modifications, by the best practical propagators everywhere at the present day.

Stocks are of all ages from a yearling seedling to a tree forty or fifty years old; but of whatever age, they should be sound and healthy. Nursery stocks will be more particularly spoken of in the proper place.

Scions are generally shoots of the previous year's growth. Rarely those bearing fruit buds are used for the purpose of experiment, but in such cases only. They should be cut in the autumn after the fall of the leaf, or in the winter, and be preserved carefully in earth till wanted for use. If intended for root-grafting early in the spring in the house, it will be sufficient to bury their lower ends in earth, in a cool, dry cellar; but if wanted for out-door grafting, they should be buried in *dry sandy soil*, in a pit, on the north side of a wall or fence, and deeply covered with earth drawn up in a mound to throw off the water. They are thus kept perfectly dormant until used, and not so dry as to shrivel the bark. They should always be taken from healthy, vigorous trees *exclusively*, and be of firm, well-ripened wood. A moderate-sized shoot or scion, if well matured and sound, is much better than one as thick as a man's finger, *pithy* and unripe. People are by no means so careful and discriminating in this respect as they ought to be. Half of the maladies of trees originate in negligent and vicious systems of propagation. The implements used in grafting are the *grafting-knife*, *saw*, and *chisel* (see implements). In whip-grafting or splice-grafting, the stocks being small require the knife only, or not more than the knife and chisel. It is always better to have two knives—one to prune and do the rough work, and the other to prepare the scion. *Grafting composition* is prepared in various ways. *Rosin*, *beeswax*, and *tallow*, in about equal parts, answer very well. Lately, however, we have found it better to use more rosin and less beeswax and tallow; thus, to two pounds of rosin we add one and one fourth pounds of beeswax, and three fourths of a pound of tallow. For whip-grafting on the root, and small trees in the nursery, we use cloth saturated with this composition, instead of the composition itself, and find it more convenient and expeditious. If we have no old calico, we buy a very thin article, at about four cents per yard. This we tear into narrow strips, roll into balls, and then soak in the liquid composition until every pore of the cloth is filled with it. The person who applies it to the grafts takes it from these balls, tears it in pieces the length and breadth required by the size of the stock, and two or three turns of it around the graft secure it completely. This thin cloth soon decays, and yields to the enlargement of the parts it encloses. We have tried tow, paper, and other materials, but find this the best. Having the scions, implements, and composition in readiness, the work is performed as follows:

Whip-Grafting on the Root.—For this purpose, seedling stocks are generally used, one or two years old, varying from *one fourth* to *three eighths* of an inch in diameter. The graft is always made at the collar, and, therefore, the stems of the plants are cut off at that point; the small tap-roots and any cumbrous fibres are removed, leaving them about four inches in length (fig. 71); they are then washed clean, and are ready for the operation. The grafter then makes a smooth, even, sloping cut, an inch long, upwards on the collar of the root, A; and in the centre of this cut, he makes a slit or tongue, B, downwards. The scion, which should be three or four inches long (fig. 72), is cut on the lower end with a sloping cut downwards, and similar in all respects to that made on the stock; a slit, or tongue, is made in it upwards, B, corresponding, also, with that on the stock; and they are then neatly fitted together, the tongue of the one within the other (A, fig. 73), and the

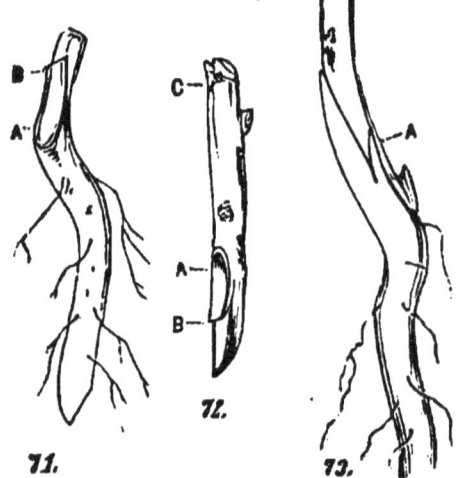

FIGS. 71 to 73, ROOT GRAFTING.

Fig. 71. the root. A, the sloping cut B, the tongue
Fig. 72. the scion. A, the sloping cut. B, tongue. C bud at top. *Fig.* 73, the union of scion and stock.

inner barks of both placed in close and perfect contact, at

least on one side. The fit should be so complete as to sit close and firm in all parts. The person who applies the wax, takes a narrow strip of the cloth described, and wraps it firmly around, covering the parts united. A man and boy can graft of these twelve to fifteen hundred per day, and by a special effort two thousand. When the grafting is thus performed, the grafted plants are put away as closely as they can be packed in small boxes, with sandy earth among the roots, and deposited either in a cold cellar or in a dry place out of doors, where frost cannot penetrate to the roots, until planting time in spring.

Whip Grafting on small trees, standing in the open ground, is performed in precisely the same manner, the oblique or sloping cut and tongue, corresponding in stock and graft, fitting into each other with precision, and the inner bark of both, at least on one side, placed in close contact. Stocks an inch in diameter can be grafted in this way. Either the cloth or the liquid composition may be applied, the latter put on with a brush. For all moderate sized stocks the cloth is preferable. In cold weather, a small furnace can be kept at hand to keep the composition in working order.

Cleft Grafting is practised on trees or branches too large for whip grafting, say from an inch in diameter upwards. In this case, the scion is cut precisely in the form of a wedge (fig. 74). The part cut for insertion in the stock, should be about an inch or an inch and a half long, with a bud (*A*) at the shoulder, where it is to rest on the stock; this bud hastens the union of the parts, in the same way as a bud at the base of a cutting, set in the earth, hastens and facilitates the emission of roots: the outer edge should also be somewhat thicker than the inner. A sloping cut (*A*, fig 75) is then made on the stock, an inch and a half long, another cut (*B*) is made *across* this cut, about half way down, as at point *B*, the stock is split on one side of the pith, by laying the chisel on the horizontal surface, and striking lightly with a mallet; the split is kept open with the knife or chisel till the scion is inserted with the thick side out (*A*, fig. 74). Grafts of this kind heal much more rapidly than when cut at once horizontally. Very large branches are sawed horizontally off at the point to be grafted (*A*, fig. 77); the surface is then pared smooth with the knife, a split is made with the chisel, nearly in the centre, and *two* wedge-like scions inserted (*A*, *B*, fig. 78);

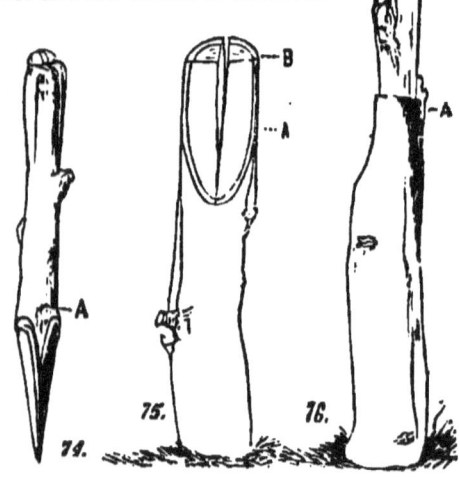

FIGS. 74 to 76, CLEFT GRAFTING.

Fig. 74, the scion prepared with a sloping cut on each side like a wedge. *A*, a bud at the shoulder. *Fig.* 75, the stock cut and split. *A*, the sloping cut. *B*, the horizontal cut. *Fig.* 76, the scion inserted in the stock.

if both grow, and they are afterwards too close, one can be cut away. Another mode of grafting such large stocks, or branches, is to cut them off horizontally, as above, and pare them smooth with the knife; then cut the scion on *one* side, about an inch and a half long, making a shoulder at the top, then raise the bark from the stock with the handle of a budding knife, and insert the scion between the bark and wood, and apply the composition the same as in the others, all over the cut part. Two or three scions may be put in each. The principal objection to this mode is, that the grafts, if they grow rapidly, are apt to be blown off before they have united strongly to the stock.

The great points to observe *always* are, to have sharp instruments that will make smooth clean cuts, to have placed in perfect contact the inner barks of scion and stock, and the whole cut surface, and every portion of the split perfectly covered with the composition, to exclude air and water. The scion should always be cut close to a bud at the point (*c*, fig. 71), and have a bud at the shoulder, or point of union with the stock (*A*, fig. 73).

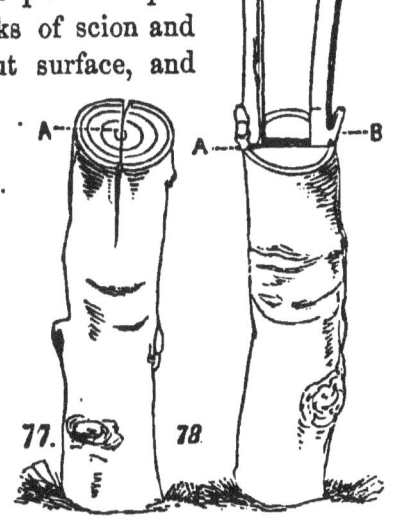

77 and 78, cleft grafting, large trees or branches. 77, the stock cut horizontally at *A*. 78, the same, with two scions inserted.

In grafting the heads of large trees, it is not convenient to use the composition in a melted state, to be put on with the brush, and the large cut surfaces cannot well be covered with the cloth; it is therefore better to use the composition in such a state that it can be put on with the hands. A very small quantity of brick dust may be advantageously mixed with it, when intended for this purpose, to prevent its being melted by the sun.

Double Working.—When we graft or bud a tree already budded or grafted, we call it "double worked." Certain very important advantages are gained by it. Some varieties are of such feeble growth, that it is impossible to make good trees of them in the ordinary way of working on common stocks. In such cases, we use worked trees of strong growing sorts as stocks for them.

Many varieties of the pear do not unite well with the quince stock; we therefore bud other varieties of strong growth that do succeed, and use them for stocks to work the others on. By this means we are enabled to possess dwarf trees of many varieties, that we could not otherwise have in that form. We have fruited the Dix in two years by double working on the quince, when otherwise it would have taken not less than seven. A great many improvements may be effected, not only in the form and growth of trees, but in the quality of the fruit, by double working. Very few experiments have yet been made on the subject in this country, except from necessity; but the general interest now felt on all matters pertaining to fruit tree culture. cannot fail to direct attention to this and similar matters that have heretofore, in a great measure, been overlooked.

CHAPTER V.

PRUNING—ITS PRINCIPLES AND PRACTICE.

This is one of the most important operations connected with the management of trees. From the removal of the seedling plant from the seed bed, through all its successive stages of growth and maturity, pruning, to some extent, and for some purpose, is necessary. It may, therefore, be reasonably presumed, that no one is capable of managing trees successfully, and especially those conducted under certain forms, more or less opposed to nature, without knowing well *how to prune, what to prune,* and *when to prune.* This knowledge can only be acquired by a careful study of the structure of trees, because the pruning applied to a tree must (aside from the general principles on which all pruning depends) be adapted to its particular habits of growth and mode of bearing its fruit. It is in view of this fact that the chapter on the structure and mode of formation of the different parts of fruit trees has been given in the first part of this treatise, that it may form the basis of this branch of culture.

The idea that our bright American sun and clear atmosphere render pruning an almost unnecessary operation, has not only been inculcated by horticultural writers, but has been acted upon in practice to such an extent that more than three fourths of all the bearing fruit trees in the country, at this moment, are either lean, misshaped skeletons, or the heads are perfect masses of wood, unable to yield more than one bushel of fruit in ten, well matured, colored, and ripened.

This is actually the case even in what may be called, in comparison, well managed orchards. Look at the difference between the fruits produced on young and old trees. The former are open, the fruits are exposed to the sun, and, therefore, they are not only large and perfect, but their skins are smooth and brilliant, as though they were painted and polished. This ought to teach us something about pruning; but this is only one point. We prune one portion of a tree to reduce its vigor, and to favor the growth of another and weaker part. We prune a stem, a branch, or a shoot to produce ramifications of these parts, and thus change or modify the form of the whole tree. We prune to induce fruitfulness, and to diminish it. We prune in the growing as well as in the dormant season; and, finally, we prune both roots and branches. Thus we see that pruning is applied to all parts of the tree, at all seasons, and to produce the most opposite results.

It appears necessary to treat of pruning under each of these circumstances separately.

1st. *Pruning to Direct the Growth from one Part of a Tree to another.*—The first period in the existence and growth of a tree in which this becomes necessary, is in the nursery. Those who have had any experience in tree culture, have observed that young trees in nursery rows have a tendency to increase in height without acquiring a well-proportioned increase in diameter. In certain cases, this want of proportion becomes so great, that the tree bends under its own weight; and hence, it is necessary to resort to some method of propping it up. This condition is attributable to several causes. First, the absence of a sufficient amount of air and light around the stem, to enable the leaves on it to fulfil their functions properly. It has been shown that the formation of new wood depends upon the elaborating process carried on in the leaves, and that this process can be maintained only in a free exposure to the sun and air. This being the case, it is obvious that any part of the tree excluded from the action of these agents, cannot keep pace in growth with other parts to which they have full access. In nursery rows, as trees are usually planted, the stems, after the first year's growth, are, to a great extent, excluded from the light, consequently the buds and leaves on them cannot perform their parts in the creation of new wood. The top of the tree, however, is fully exposed, and, consequently, it makes a rapid growth towards the free air and light. When this is continued for two or three years in succession, the tree becomes top-heavy; the quantity of woody fibre at the top is as great as, and it may be greater than, at the bottom; and hence it bends under its own weight.

2d. *The Tendency of the Sap to the Growing Points at the Top of the Tree.*—Growth is always the most active and vigorous, when trees are in a natural condition, at the newly-formed parts. The young buds are the most excitable, and the more direct their communication with the roots, the more rapid will be their growth. Hence it is that a yearling tree furnished with fifteen to twenty buds or more, from its base to its top, frequently produces a shoot from its terminal bud only, and seldom more than three or four shoots from the whole number of buds, and these at the top. This natural tendency, and the exclusion of light from the stems of nursery trees, by their closeness to one another, are the chief causes of weak and crooked trees, to counteract which we resort to *pruning.*

In "*heading down*" *a young tree*, we cut away one third or one half of the length of the stem, and this removes the actively growing parts; the sap must then find new channels. Its whole force is directed to the buds that were before dormant, they are excited into growth, and produce new wood and leaves; these send down new layers of woody fibre on the old stem, and it increases rapidly in diameter, so that by the time it has attained its former height, the base is two or three times as thick as the top, and possesses sufficient strength to maintain an erect position.

Maintaining an equal growth among the branches of a tree is conducted on the same principle. Branches that are more favorably placed than others, appropriate more than their due proportion of the sap, and grow too vigorously, are checked, by removing more or less of their growing points; this lessens the flow of sap to that point, and it naturally takes its course to the growing parts of the weaker branches that were left entire, and thus a balance is restored.

Pruning to renew the Growth of Stunted Trees.—It frequently happens that trees, from certain causes, become stunted, and almost cease to grow; the sap vessels become contracted, and every part assumes a comparatively dormant condition. In such cases they are cut back, the number of their buds and leaves is reduced, the whole force of the sap is made to act upon the small number remaining, and enables them to produce vigorous young shoots; these send down new woody matter to the stem, new roots are also formed, and thus the whole tree is renewed and invigorated.

Pruning to induce Fruitfulness.—This is conducted on the principle that whatever is favorable to rapid, vigorous growth, is unfavorable to the immediate production of fruit. Hence the object in view must be to check growth and impede the circulation of the sap, just the opposite of pruning to renew growth. The only period at which this pruning can be performed, is after vegetation has commenced. If a tree is severely pruned immediately after it has put forth its leaves, it receives such a check as to be unable to produce a vigorous growth the same season; the sap is impeded in its circulation, and the result is that a large number of the young shoots that would have made vigorous wood branches, had they not been checked, assume the character of fruit spurs and branches. *Pinching* is the principal mode of pruning to promote fruitfulness, and will be explained hereafter. It depends upon the above principle, of impeding the circulation of the sap and checking growth.

Pruning to diminish fruitfulness, is conducted on the same principle as that to renew growth, for this, in fact, is the object.

Pruning the Roots.—This is practised as well to promote fruitfulness, as to lessen the dimensions of trees. The roots, as has been shown, are the organs that absorb from the ground the principal food of the tree, and in proportion to their number, size, and activity, other things being equal, are the vigor and growth of the stem and branches. Hence when a tree is deprived of a certain portion of its roots, its supply of food from the soil is lessened, growth is checked, the sap moves slowly in its channels, is better elaborated in the leaves, and the young branches and buds begin to assume a fruitful character.

Roots are also pruned to prevent them from penetrating too deeply into the earth, and induce the formation of lateral roots near the surface, similar to the cutting back of a stem to produce lateral branches; the principle is the same.

Pruning at the time of Transplanting.—This is performed, not only to remove bruised and broken roots and branches, but to restore the tree to a proper balance. As trees are ordinarily taken from the ground, the roots are bruised, broken, or mutilated, to a greater or less extent. This obviously destroys the natural balance or proportion that existed between the roots and stem, and in such a condition the tree is unable to grow. The demand upon the roots must therefore be lessened, by reducing the stem and branches in length or number, or both; and the more the roots have suffered, the greater must be the reduction of the stem and branches, to bring them to a corresponding condition.

PRUNING MECHANICALLY CONSIDERED.

Having now treated of the principles on which pruning depends, it remains to speak of its mechanical execution; for it is not only necessary to know what and why, but *how* to prune. Theory is only useful as it serves to guide in practice.

1st. *Pruning Stems* or *Branches.*—The great point to be observed in making incisions on the stems and branches of trees, is to provide for the speedy and perfect healing of the wounds or cut surfaces. In removing a portion of a branch or stem, if we cut between two joints, and thus leave a portion of wood above the bud intended to be cut to, as in fig. 79, this wood dies, and we have the trouble of another pruning to remove it. If we cut too close to the bud, and thus remove a portion of the wood with which it is connected, as in fig. 80, the bud will either die or disappoint us by producing a very feeble growth. The proper way is to take the branch to be operated on in the left hand, place the edge of the knife on it, opposite the lower part of the bud to be cut to, and then make a firm, quick, smooth draw-cut, sloping upwards, so that the knife will come out on a level with the point of the bud, as in fig. 81. In soft-wooded, pithy trees, like the grape vine, for example, half an inch of wood ought to be left above the bud. The cut should also be made as much as possible on the lower side of the branch to prevent rain from lodging in the centre. The position of the bud cut to, is also worthy of consideration in pruning, to produce or modify certain

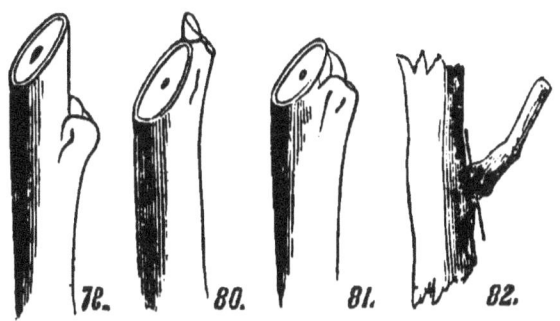

Figs. 79 to 82, PRUNING.

Fig. 79, cutting too far above the bud. *Fig.* 80, cutting too close. *Fig.* 81, the cut as it should be. *Fig.* 82, removal of a branch, the cross line indicating the proper place for the cut.

forms. When we wish the new shoot of a lateral branch to take, as much as possible, an *upright* direction, we prune to a bud on the *inside;* and if we wish it to *spread*, we choose one on the *outside*. In the annual suppression, of cutting back young trees, to form a stem or side branches, the bud selected to form the leader is chosen on *opposite sides every successive year*, in order to maintain the growth in a straight line. If cut every year to a bud on the same side, it would, in two or three seasons, show an inclination to that side injurious to the symmetry of the tree.

The Removal of Large Branches, where they are to be entirely separated from the tree, is often very clumsily performed. In orchards, it is not at all uncommon to see them chopped off with a common axe; and even in gardens there seem to be few persons who either know how, or take the proper care in this matter. They are either cut so that a portion of the base of the branch remains, and sends out vigorous shoots, defeating the objects of the pruning, or they are cut so close that a portion of the wood of the main branch or stem is taken with them, and a wound made that years are required to heal up. Both these extremes ought to be avoided.

The surface of the cut made by the removal of a branch should in no case be larger than the base of the branch. Where a branch is united to another, or to the main stem, we notice both above and below the point of union, a small projection or shoulder, as at the cross line in fig. 82. The knife must enter just below that shoulder, and, by being drawn upwards in a straight line, the base is so completely removed that no shoots can be produced there; and yet the cut surface on the stem is no larger than the base of the branch. When the saw is used, the surface of the cut should be pared smooth with the knife, to prevent water lodging on it, and facilitate the healing of the wound.

2d. *Pruning the Roots.*—This is performed by opening a trench around the tree, just at the extremities of the roots : the distance from the tree will, therefore, depend on its size, and the spreading characters of the roots. The trench should be the width of a common garden spade, and deep enough to admit of an inspection of all the roots of the tree. If the lateral roots are to be shortened, this is done first. The knife should be placed on the lower side of the root, and the part separated with a clean draw-cut, such as would be performed on a branch. If the tree has vertical, or tap roots, they are most easily operated on with a sharp spade, prepared and kept for the purpose. A smart stroke with such a spade, in as nearly a horizontal direction as possible, will separate a pretty strong root. The extent to which root pruning may be performed, depends on the character of the species, the condition of the tree as regards growth, and the object aimed at. Those practising it for the first time, should go to work with great caution. It will be better to operate too lightly than too severely. As regards the season, it may be performed either at the end of the first growth, in July or August, or in the autumn or winter, when vegetation is quite suspended. We have operated on cherry trees with complete success in August, in a dry time, when little growth was going on. At this season, a copious watering should be given after the pruning is performed.

Implements of pruning, and the mode of using them, will be treated of in the chapter on implements, to be given hereafter.

The Season for Pruning.—We are not permitted to be very definite on this point. The climate, the nature of the species, etc., control the period of pruning to a great extent. In the south, what we term the winter pruning—that performed during the dormant season—may be done very soon after the fall of the leaf. In the north, it is deferred to February, March, and even April. In western New York, we prune *apples*, *pears*, and other hardy fruits, as soon as our severe frosts are over—say the latter end of February and beginning of March. If pruned sooner, the ends of the shoots are liable to be injured, and the terminal bud so weakened as not to fulfil its purposes. Besides, the wounds do not heal well.

The *peach* we prune just as the buds begin to swell. The fruit and leaf buds are then easily distinguished from one another, and the objects of the pruning are accomplished with more precision.

Grapes may be pruned any time in the winter, as a portion of wood is always left above the bud. *Gooseberries* and *currants* also, any time in winter. The stone fruits should always be lightly pruned, because severe amputations almost invariably produce the gum. Where it is absolutely necessary in the spring, the wound should be coated with grafting composition, or with that recommended by Mr. Downing: "Alcohol, with sufficient gum shellac dissolved in it, to make a liquid of the consistence of paint, to be put on with a brush."

This excludes air, and is not affected by changes of weather.

Pinching is a sort of anticipated pruning, practised upon the young growing shoots, intended to promote a uniform circulation of the sap, and thus regulate the growth, and also to induce fruitfulness.

1st. *To Regulate the Growth.*—In the management of trees, this is an operation of great importance, as it obviates the necessity of heavy amputations being made at the winter or spring pruning. Instead of allowing certain superfluous or misplaced shoots to acquire their full development at the expense of other parts, we pinch them early, and give to the necessary parts, or branches of the tree, the nutriment which they would have appropriated, if allowed to remain. In this way, we are able to obtain results in one season, that two or more would be required for, if we depended wholly on the winter pruning. We will suppose, for an example, the case of a young nursery tree in the second year, intended for a standard. In ordinary cases, the terminal bud, either the natural one or that pruned to, is developed into the leading shoot or stem, and a greater or less number of buds below it produce branches; and it frequently happens that some of these, if not pinched, acquire so much vigor as to injure the leader and produce a consequent deformity in the tree. Fig 83 (*A*) represents a case of this kind, which is very common, and too often neglected. The shoots, *a, a*, ought to have been pinched the moment they began to exhibit a disposition to outgrow the leader. There are other cases still worse than this, familiar to all tree growers; for instance, where a strong shoot is produced on the middle or lower

PRUNING. 93

part of the stem, attracting an undue proportion of the sap, thus contracting the growth of all other parts, and giving the young tree a deformed character. All such

Fig. 83 (A). Fig. 83 (B).

Fig. 83, A, head of a young tree; B, the leader; a, a, vigorous shoots below it, that ought to have been pinched. Fig. 83, B, a branch of the pear, twice cut back with the lateral shoots pinched; a, a, the first section; c, c, c, the second; b, and d, d, shoots pinched close to favor the leader, and those below them.

shoots as these should be nipped early, the moment their character is apparent, and thus a year's growth nearly will be saved to the tree, and its proper form and proportions be preserved. In conducting young trees for pyramids, the constant and careful application of pinching is absolutely necessary, for in them we must have the lower branches always the strongest and longest, and it is only by operating on the shoots, in their earliest stages of growth, that we can fully attain this end; for the strongest shoots do not always grow at the desired point, but by timely attention they are perfectly within our control. The various accidents and circumstances to which young trees are subject, give rise, in a multitude of cases, to an unequal distribution of the sap in their different parts, and this produces, to a greater or less extent, deformity of growth. This, at once, shows the necessity for pinching, to check the strong and favor the weak.

Pinching to promote Fruitfulness.—Those who have never practised this, or observed its results, may have seen, if experienced in tree growing, that a shoot of which the point was broken, bruised, or otherwise injured, during the growing season, frequently becomes a fruit branch, either during the same or the following season; and this, especially if situated in the interior of the tree, or on the older and lower parts of the branches. The check given to the extension of the shoot concentrates the sap in the part remaining; and, unless the check has been given very early in the season, or the growth very vigorous in the tree, so that the buds will break and form shoots, they are certain to prepare for the production of fruit. It is on this principle of checking the growth, and concentrating the sap in the pinched shoot, that pinching to induce fruitfulness is performed; and its efficiency may be estimated from the fact, that trees on which it has been practised, have borne fruit four or five, and perhaps seven years, sooner than they would have done without it.

It is a most useful operation in the case of vigorous growing and tardy bearing sorts. The best illustration, on a large scale in this country, is the specimen plantation of pear trees of Messrs. Hovey & Co., of Boston. A large number of these are pyramidal in form, and on pear stocks, very beautiful trees, indeed the best specimens of the kind in any American nursery, and though, now in 1850, only 7 years old (the oldest), yet they have as a general thing produced fruit, and many of them for 2 or 3 years past. This result has been obtained by pinching, which has been regularly, but not to the fullest extent, practised upon them every season. The *mode of performing it*, is to pinch off the end of the shoot with the finger and thumb; if a small portion of the remaining part be bruised, no matter, it offers a greater check than if a clean cut were made, as in pruning to a bud; and in the general winter or spring pruning which follows, the bruised parts can be cleanly separated. The *time to perform it* depends wholly on circumstances. If the object be to regulate growth, then the time to do it is, when the tendency to undue or ill-proportioned growth is first observable, and this will be from the time the young shoots are two to three inches long and upwards. The particular season of the year or day of the month will, of course, depend upon the earliness or lateness of the season, and on the soil and situation as well as on the habits of growth of the species or variety to be operated on. The true way is to be always on the watch. If the object be to induce fruitfulness, the length which the shoots should attain before being pinched, depends upon the nature or mode of growth and bearing of the species, and will be more definitely treated under the head of "The Pruning of Trees," hereafter, the object now being merely to indicate general principles and modes of operating. To illustrate this, let us suppose the lateral branch of a pear tree, (fig. 83, *B*). This was cut back the first time to *b*, and below that point five shoots were produced, none of which were needed for branches. We, therefore, pinched them in June, when about three inches long or thereabouts, and the result is, they are now fruit branches. The same branch was cut back the second time to d, d, and on that section seven shoots were produced that were not needed in the form of the tree, and were consequently pinched, and will become fruit branches. At the points b, and d, d, are small spurs, the base of shoots that have been pinched close to favor the growth of the leader, as well as the development of the shoots below. Without pinching it would have been impossible to obtain such results in this branch in the same time.

M. Dubreuil, formerly Professor of Arboriculture in the Garden of Plants at Rouen, in France, sums up the general principles of pruning as follows. (I may remark here, that in 1849, I visited the Rouen garden, and found M. Dubreuil's theory and practice beautifully illustrated on the trees in his charge. My visit was made at the time of his practical lectures, and I was able to examine the whole with the most satisfactory minuteness. The trees there, under all forms, and embracing all the hardy species of fruits, were the best that I anywhere found, not even excepting the much admired and famous pyramidal pear trees of M. Cappe, at Paris. They were not only perfect in form, but as regards *vigor* and *fruitfulness*, in the most admirable condition.) He says:

" The theory of the pruning of fruit trees rests on the following six general principles:

" 1. *The vigor of a tree, subjected to pruning, depends, in a great measure, on the equal distribution of sap in all its branches.*

" In fruit trees abandoned to themselves, the sap is equally distributed in the different parts without any other aid than nature, because the tree assumes the form most in harmony with the natural tendency of the sap.*

---

* This is not in all cases true. Peach trees, we know, left to themselves, exhibit a very striking example of the unequal distribution of the sap. The ends of the branches attract nearly the whole, leaving the lateral shoots and

"But in those submitted to pruning, it is different; the forms imposed on them, such as espalier, pyramid, vase, &c., change more or less the normal direction of the sap, and prevent it from taking the form proper to its species. Thus nearly all the forms given to trees require the development of ramifications more or less numerous, and of greater or less dimensions at the base of the stem. And, as the sap tends by preference towards the summit of the tree, it happens that, unless great care be taken, the branches at the base become feeble, and finally dry up, and the form intended to be obtained disappears, to be replaced by the natural form, that is a stem or a trunk with a branching head. It is then indispensable, if we wish to preserve the form we impose upon trees, to employ certain means, by the aid of which the natural direction of the sap can be changed and directed towards the points where we wish to obtain the most vigorous growth. To do this we must arrest vegetation in the parts to which the sap is carried in too great abundance, and on the contrary favor the parts that do not receive enough. To accomplish this the following means must be successively employed.

"1. *Prune the branches of the most vigorous parts very short, and those of the weak parts long.* We know that the sap is attracted by the leaves. The removal of a large number of wood-buds from the vigorous parts, deprives these parts of the leaves which these buds would have produced; consequently the sap is attracted there in less quantities, and the growth thereby diminished. The feeble parts being pruned long, present a great number of buds, which produce a large surface of leaves, and these attract the sap and acquire a vigorous growth. This principle holds good in all trees, under whatever form they may be conducted.

"2. *Leave a large quantity of fruit on the strong part, and remove the whole, or greater part, from the feeble.* We know already that the fruit has the property of attracting to it the sap from the roots, and of employing it entirely to its own growth.

lower parts to die out. In other species, similar instances might be quoted, and as a general thing, the proposition is unsound, except in a comparative sense.

The necessary result of this is, what we are about to point out, viz., that all the sap which arrives in the strong parts, will be absorbed by the fruits, and the wood there, in consequence, will make but little growth, while on the feeble part, deprived of fruits, the sap will all be appropriated by the growing parts, and they will increase in size and strength.

" 3. *Bend the strong parts and keep the weak erect.* The more erect the branches and stem are, the greater will be the flow of sap to the growing parts; hence, the feeble parts being erect, attract much more sap than the strong parts inclined, and, consequently, make a more vigorous growth, and soon recover their balance. This remedy is more especially applied to espalier trees.

" 4. *Remove from the vigorous parts the superfluous shoots as early in the season as possible, and from the feeble parts as late as possible.* The fewer the number of young shoots there are on a branch, the fewer there are of leaves, and consequently the less is the sap attracted there. Hence, in leaving the young shoots on the feeble part, their leaves attract the sap there, and induce a vigorous growth.

" 5. *Pinch early the soft extremities of the shoots on the vigorous parts, and as late as possible on the feeble parts, excepting always any shoots which may be too vigorous for their position.* By thus pinching early the strong part, the flow of sap to that point is checked, and naturally turns to the growing parts that have not been pinched; this remedy is applicable to trees in all forms.

" 6. *Lay in the strong shoots on the trellis early, and leave the feeble parts loose as long as possible.* Laying in the strong parts obstructs the circulation of the sap in them, and consequently favors the weak parts that are loose. This is only applicable to espaliers.

" 7. *In espalier trees, giving the feeble parts the benefit of the light, and confining the strong parts more in the shade, restores a balance,* for light is the agent which enables leaves to perform their functions and their action on the roots, and the parts receiving the greatest proportion of it acquire the most vigorous development.

2. "*The sap acts with greater force and produces more vigorous growth on a branch or shoot pruned short, than on one pruned long.*" This is easily explained. The sap acting on two buds must evidently produce a greater development of wood on them, than if it were divided between fifteen or twenty buds.

"It follows from this, that if we wish to obtain wood branches, we prune short, for vigorous shoots produce few fruit buds. On the contrary, if we wish to obtain fruit branches, we prune long, because the most slender or feeble shoots are the most disposed to fruit.

"Another application of this principle is to prune short for a year or two, such trees or parts as have become enfeebled by overbearing. (This principle deserves especial attention, as its application is of great importance.)

3. "*The sap tending always to the extremities of the shoots causes the terminal bud to push with greater vigor than the laterals.*" According to this principle, when we wish a prolongment of a stem or branch, we should prune to a vigorous wood bud, and leave no production that can interfere with the action of the sap on it.

4. "*The more the sap is obstructed in its circulation, the more likely it will be to produce fruit buds.*" This principle is founded on a fact to which we have already had occasion to refer, viz.— that the sap circulating slowly is subjected to a more complete elaboration in the tissues of the tree, and becomes better adapted to the formation of fruit buds.

"This principle can be applied to produce the following result: When we wish to produce fruit buds on a branch, we prevent a free circulation of the sap by bending the branches, or by making annular or circular incisions on it; and on the contrary, when we wish to change a fruit branch into a wood branch, we give it a vertical position, or prune it to two or three buds, on which we concentrate the action of the sap and thus induce their vigorous development.

5. "*The leaves serve to prepare the sap absorbed by the roots for the nourishment of the tree, and aid the formation of buds on the shoots. All trees, therefore, deprived of their leaves are liable to perish.*" This principle shows how dangerous it is to remove a

large quantity of leaves from trees, under the pretext of aiding the growth or ripening of fruits, for the leaves are the nourishing organs, and the trees deprived of them cannot continue to grow, neither can the fruit; and the branches so stripped will have feeble, ill-formed buds, which will, the following year, produce a weak and sickly growth.

6. "*Where the buds of any shoot or branch do not develope before the age of two years, they can only be forced into activity by a very close pruning, and in some cases, as the peach, this even will often fail.*" This last principle shows the importance of pruning the main branches of espaliers particularly, so as to ensure the development of the buds of their successive sections, and to preserve well the side shoots thus produced, for without this, the interior of the tree will become naked and unproductive, and a remedy will be very difficult."

If these principles and practices of pruning be carefully studied in connection with the habits of growth and bearing of the different fruit trees, pruning will be comparatively an easy matter. The mode of obtaining any particular form or character cannot fail to be perfectly plain and simple; yet no one need hope to accomplish, in all things, the precise results aimed at, for even the most skilful operator is sometimes disappointed: but those who give constant attention to their trees, will always discover a failure in time to apply a remedy.

I insist upon it, because I have been taught it by most abundant experience, that the most unremitting watchfulness is necessary in conducting trees in particular forms. It is not, by any means, *labor* that is required, but attention that the most delicate hand can perform, fifteen or twenty minutes at a time, say three times a week during active growth, will be sufficient to examine every shoot on a moderate collection of garden trees; for the eye very soon becomes trained so well to the work, that a glance at a tree will detect the parts that are either too strong or too weak, or that in any way require attention. This is one of the most interesting features in the management of garden trees. We are never allowed to forget them. From day to day they require some attention, and offer some new point of interest that attracts us to them, and augments our solicitude for their prosperity, until it actually grows into enthusiasm.

PART II.

THE NURSERY

THE NURSERY.

CHAPTER I.

SECTION 1.—SOIL, SITUATION, ETC.

It is not a part of the design of this treatise to give anything like a full exposition of nursery operations; for this would, in itself, be a subject sufficiently extensive to form a volume; but as all fruit growers should possess at least some knowledge of nursery management, it seems quite necessary that the more important points should be noticed.

1st. *The Soil, as to Dryness.*—For a fruit tree nursery the soil must be *perfectly dry*, both above and below. In damp, springy soils, or where the subsoil is so compact as not to admit of the surface water passing off immediately, trees do not thrive, the roots are destitute of fibres, the wood is watery and delicate, and where frosts are severe the trees are cast out of the ground by the expansion of the water with which the soil is filled. We have known of a single instance in which several thousand dollars were lost by planting a pear nursery on a soil imperfectly drained. The plants grew finely the first season, were budded, the buds had taken, and in the autumn all looked prosperous; but the autumn rains filled the soil with water, the situation was low and level, and the subsoil compact, so that the water could not possibly get away. The consequence was, the roots decayed, the plants were cast out of the ground, and the injury was so great and so general that the whole plantation had to be taken up. This ground was then thoroughly drained, and is now as good a pear soil as can be found—a stock of beautiful trees standing on it at the present time. This single instance illustrates the importance of a dry soil, as well as twenty would. We frequently find that in the same row of trees, if there happens to be a low, damp spot, the trees in it have no fibrous roots, and are altogether inferior to those on the adjacent dry ground.

2d. *Depth.*—As a general thing, the soil of a nursery should be a foot to eighteen inches deep; but all trees do not require the same depth. Those (such as the pear) whose roots *descend* more than they *spread*, require the deepest soil. The best quality of nursery trees are grown on common farming land, twice ploughed with the common and subsoil ploughs, one following the other, as described in the chapter on soils. This gives depth enough for all ordinary purposes.

3d. *Texture.*—A soil of medium texture between the heavy and the light, is, on the whole, the most advantageous, as being the best adapted to general purposes. A good friable loam, with a gravelly subsoil, or a mixture of sand, gravel, and clay, that will allow water to pass off freely, and yet not too fast, will be found suitable for almost any species; and one great advantage of such a soil is, that it admits of rotation in crops.

4th. *Quality.*—For the growth of young fruit trees, a soil should be in such a condition as to furnish a sufficient supply of nutriment to ensure a vigorous and robust growth; but it may be too rich, and produce rank wood that will not mature properly, and be unable to withstand the change of climate or soil consequent upon transplanting. Where manures are used, they should be well decomposed; fresh warm manures excite trees into a very rapid growth, but the wood is watery and feeble. A dry soil of moderate richness produces hardy trees, their wood is firm, the buds plump and close together, and the parts well proportioned.

5. *Laying out.*--Where the nursery is of considerable extent, the ground should be laid out and arranged in square or rectangular plots of convenient size, and be intersected with walks. One portion should be set apart for the propagation of stocks from layers, another for cuttings, another for seeds, &c. In setting apart ground for the different kinds of trees, if there be a choice, the pear should have the deepest and best, the plum the most compact or clayey, the peach, apricot, cherry, &c., the lightest and dryest.

6. *Exposure.*—Nursery ground for fruit trees should be well elevated, but not fully exposed to the prevailing high winds, as the young trees are apt to be broken off during the first year's growth if not kept well tied up to stakes. In our section we find it very advantageous to have some protection from the west winds especially, though we sometimes have a south wind quite destructive in exposed places to the young buds. Situations where snow is liable to drift into, should be avoided, in sections where heavy snow storms prevail, for sometimes vast quantities of trees are broken down in corners of fences and sheltered situations where the snow accumulates in heavy drifts.

7. *Rotation or Succession of Crops.*—This is quite as important in the management of the nursery as of the farm. Not more than one crop of one species should be planted on the same ground; and those of the most opposite character should follow one another. Where one species is grown on the same ground for eight or ten years, it is found by experience that even the most liberal manuring fails to produce such fine, sound, healthy, and vigorous trees as new ground without manure. Where land is scarce, and it is necessary to use the same ground for the

same kind of trees, it should at least be allowed one season's rest, and be well supplied with such material as the trees to be grown in it require in the largest quantities, or in which the soil is found to be most deficient.

SECTION 2.—DESCRIPTION AND PROPAGATION OF STOCKS.

This branch of the subject is of such importance, and involves so many considerations, that it seems to be more methodical to treat it separate from subsequent operations.

1st. *Stocks for the Apple.*—The principal stocks in use for the apple are the *common seedling*, or *free stock*, the *Doucain*, and the *Paradise*.

Seedlings, or *free stocks*, are ordinarily produced from seeds taken promiscuously from the cider mill in the autumn.

Preparing the Seed.—The cakes of pressed pomace are broken up, and the coarser materials, straw, &c., separated from it by means of a coarse sieve, the sifted pomace is then put into large tubs, and subjected to repeated washings until clean. The clean plump seed falls to the bottom, and the pomace and light poor seed are carried off in the washings. When fruits have been selected for the seeds, they are placed in heaps until fermentation and decay have reduced the flesh to a soft pulpy state, when they are washed in tubs, in the same manner as pomace.

Saving the Seed.—When the seed is washed out as above, it must be spread thinly on boards, and repeatedly turned over until perfectly dry, when it is put away in boxes, mixed with sand, containing a slight degree of moisture. The boxes should be well secured against vermin, and be kept in a dry, cool place, till the time of planting.

Season and Mode of Planting.—If the ground be in readiness, and perfectly dry and friable, the best time is the fall, as soon as the seeds are cleaned. At this season the pomace, seeds and all, as it comes from the press, may be planted without any washing. It should be broken up fine, so that it may be evenly distributed in the seed bed. The difficulty of doing this, is a serious objection to this mode. By taking some pains in the sowing, we raise as good stocks in this as in any other way; the decayed pulp contributes considerable nutriment to the young plants in their earliest stage of growth.

When deferred till spring, it should be done at the earliest moment that the condition of the ground will admit. When the ground is ready, a line is stretched along one side of the plot, and a drill opened with a hoe about eight or ten inches wide and three deep; the seeds are then dropped, and the fine earth drawn over them with the hoe as regular as possible, covering them about three inches deep. If some leaf mould from the woods or old decomposed manure in a fit state for spreading could be had, and a covering of an inch deep of it spread on the top of the drills, it would prevent the surface from baking or cracking, and allow the plants to come up with greater strength and regularity. Whatever depth of such a covering be used, should be deducted from the covering of common earth.

Distance to Plant.—When large quantities are raised, the drills should be three feet apart to admit of the cultivator passing between them; for the ground should be kept perfectly clean and mellow around seedlings the whole season.

After Management.—It is of great importance that they be not in any way stunted, either in first coming through the soil by a hard surface, or afterwards by weeds and lack of culture; seedlings stunted during the early stages of their growth never make vigorous, healthy stocks, and indeed should never be planted. When they appear above the surface, and are too close together, they should as soon as possible be thinned out to regular distances; for when grown up in dense masses, they are generally feeble and worthless. One hundred good vigorous stocks are worth five hundred poor ones. It is very common to see seedlings of one year larger than those of two years, under different management, and in such a case the yearlings are worth twice as much as the others. A very good plan is to thin out all the weakest plants when about four or five inches high, leaving those only of vigorous habit and large foliage.

The Doucain is a distinct species of apple; the tree is of medium size, bears small sweet fruit, and reproduces itself from seed. It is used for stocks for apple trees of medium size, *pyramids*, or *dwarf standards* for gardens. It is propagated almost exclusively from layers; see fig. 63. The plants to be propagated from are planted in a rich deep friable soil, and cut back to within four to six inches of the collar; the buds, or the part below the cut, will, during the next season, produce strong shoots; the following spring the earth is drawn up around each plant in the form of a mound, so that the whole of the stem and the base of all the shoots will be covered at least three inches deep; during that season all the shoots will produce roots, and should be separated from the mother plant or stool, as such plants are termed, in the fall. If left on till spring the frost would be likely to injure them. The stools are then dressed, the soil around them is spaded up and enriched with well decayed manure, and the following season another crop of shoots is produced, much more numerous than the first, to be treated in the same way. Every year these stool plants increase in size and in the quantity of their productions, if well treated. Another course, but not so good, is frequently pursued when stocks are scarce. The shoots are layered, by bending down as described in layering, the first season of their growth in July, and may be sufficiently rooted in the fall to be transferred to nursery rows in the spring following; a year is thus saved, but the stocks are, of course, much inferior. If earthed up in midsummer, they will be partially rooted in the autumn too, but not so well as if bent down, for the bending has a tendency to stop the sap at the point fastened to the ground, and hastens the formation of roots.

The Paradise.—This also is a distinct species of apple. The tree is of very small size, never attaining over three to four feet in height. It is used for stocks for dwarf trees or bushes that occupy but a small space in the garden. It is propagated in precisely the same manner as that described for the Doucain.

2d. *Stocks for the Pear.*—The *pear seedling* and the *quince* are the only two stocks on which the pear can be advantageously worked to any considerable extent. The mountain ash and the thorn are occasionally used for special purposes only.

Pear Seedlings.—The seeds are obtained by collecting such fruits as can be had, containing perfect seeds. Great care should be taken to gather the fruits of hardy, healthy, vigorous trees only, and the seeds should be full and plump. The seeds are separated and washed, as described for apples. They are also saved and planted in a manner similar in all respects; but in this country it is a much more difficult matter to succeed with pear seedlings than with the apple. This difficulty is owing chiefly to a species of rust or blight that attacks the leaves of the young plants, very often before they have completed their first season's growth. To obviate the difficulty which this malady presents, a vigorous growth should be obtained early in the season. New soil, or that in which trees have not been grown in before, should be selected. The autumn before planting, it should be trenched or subsoil ploughed to the depth of two feet, for the pear has long tap roots, and liberally enriched with a compost of *stable* manure, *leaf mould* or *muck*, and *wood ashes*, in about equal parts : four inches deep of this spread over the surface before ploughing, will be sufficient for any ordinary soil. Lime should also be given liberally, unless the soil be naturally and strongly calcareous. A soil prepared thus in the fall, will require another ploughing or spading in the spring, to mix all the materials properly with the soil, and fit it for the seeds. Where large quantities are grown, the drills may be the same distance apart as that recommended for apples, *three feet;* but if only a few, twelve to eighteen inches will be sufficient, as the cleaning can be done with the hoe. The seeds should be scattered thinly, that every plant may have sufficient space without any thinning. From time to time we find regular recipes given for raising pear seedlings, with the same precision that pudding recipes are given in the cook books. *Bone dust, blacksmiths' cinders, muck, lime, wood ashes*, and half a dozen other things, are recommended to be compounded in pecks and half pecks, all with a view to remedy the rust or leaf blight that no man can say originates in any defect of the soil. The cause may be in the atmosphere, or it may be an insect, or it may be something else, for aught anybody yet knows to the contrary. The end to aim at, as before remarked, is to get good growth, say eighteen to twenty inches in height, and stout in proportion, before the first of August. This can be done in any deeply-trenched, fresh soil, well prepared and manured as described above. During the past season, a lot of very fine seedling pears were raised in fresh, new soil, in Ontario county; their foliage was quite fresh when the frosts came, and they had received no special manuring either. Pear seedlings should always be taken up in the fall, after the first season's growth, the largest selected for transplanting into the nursery, and the smaller to be put into beds, to remain another season.

Quince Stocks are propagated with considerable success by cuttings. These should be strong shoots, six inches to a foot long, taken off close to the old wood, and, if possible, with a small portion attached, prepared as directed in article on cuttings, early in the winter, and kept in pits two or three feet below the surface of the soil, in a dry place, till planting time in spring. They should be planted in a *light, friable, deep soil,* in rows eighteen inches to two feet apart, four to six inches apart in the row, and so deep that but a couple of buds remain above the surface. The ground should be kept clean and mellow amongst them all summer, and if the cuttings were stout and long, they will in the autumn be fit for taking up and preparing for planting into nursery rows the following spring. The best and surest method of propagating the quince stock, however, is by *layers,* as the best variety for that purpose does not strike so freely from cuttings as the common sorts. The manner of layering is that recommended for the *Doucain* and *paradise,* by earthing up. The stool plants should be set out in a fine, rich, deep border of warm, friable soil, and be about six feet apart, when designed to be permanent. As each stool, by the system recommended, can only yield a crop of plants every two years, there should be two sets, so that an annual supply may be obtained.

By the ordinary system of bending down the shoots, and slitting, or even without the slitting, a crop may be obtained every year, that is, the shoots of the current season's growth may be layered in July or August, but no such stocks can be obtained as by the earthing up and taking a crop every two years. This is the system recommended to those who want *first rate quince stocks*.

The very general lack of information in this country on the subject of quince stocks for pears has given rise to a great many misapprehensions and erroneous statements in regard to them, both by horticultural writers and others. At first it was said that the stock used by the French and imported by nurserymen here were the *Portugal*. Again, it was discovered they were nothing more than the common apple quince; consequently a multitude of the apple quinces have been worked, and sent out as "*dwarf pears*." The slow and feeble growth of this variety unfits it entirely for a stock for the pear, and only a very few varieties will form a union with it that will last over three or four years. Such trees cannot fail to give general dissatisfaction, and among people who know no better, create a prejudice against quince stocks in general. Indeed this is the cause why so much has been said about the pears on quince being so short-lived.

The truth is, that the varieties used in France are neither the Apple nor the Portugal Quince, but vigorous hybrids that have been originated there, and found to answer this purpose particularly well. The great requisite of a quince stock for the pear is a *free, vigorous* and *rapid growth*. A variety originated at the town of Angers in France, and extensively used, propagated and sold there as the *Angers Quince*, is probably the best yet known for a pear stock generally. It is a very rapid, vigorous grower making strong shoots three feet long in one season. It has large foliage resembling the Portugal. In some parts of France, as in Normandy, it is known as the *broad-leaved* There is another variety with smaller leaves, but of free vigorous growth too, almost exclusively cultivated in some districts. Several extensive nurserymen at Orleans, Paris, and elsewhere, consider it superior to the broad-leaved, and especially for very vigorous growing sorts. It is known as the small-leaved.

We have tried both extensively, and find but very little difference thus far in the results obtained. We are now engaged in experiments testing the fitness of another variety quite distinct in its character, habits of growth, &c., from all the others. It is remarkably erect, with a bushy, branching head, and roots composed almost entirely of fine fibres. Every cutting grows when other sorts are a complete failure; and a cutting made of a stout shoot set in the ground in April may be budded in September. The largest plants we have are but three years old; and judging from these, it will not attain so large a size as the Angers, but the pear seems to unite well with it, and we believe it will make an excellent stock, for free growing kinds particularly. It is yet too soon, however, to decide upon its merits in any respect, except that of being easily propagated.

The *Mountain Ash*, it is said, makes a good stock for certain varieties in very light, sandy soils, when neither the pear nor quince succeeds well. It is propagated from seed, and requires to be two years old before being worked.

The Thorn.—Seedlings of our vigorous native thorns make good stocks when about three years old; the seeds require to be in the rot heap one year before sowing. The only cases in which it can be recommended, are those in which a soil may be so wet and cold as to be unfit for the pear or quince; but it is better to improve such soils by draining, subsoil ploughing, and by the addition of suitable composts, for even the thorn will fail in giving satisfaction on a stiff, cold soil.

3d. *Stocks for the Cherry.*—The principal stocks used for the cherry are the *mazzard* for standard orchard trees, and the *mahaleb* for garden pyramids and dwarfs.

Mazzard Seedlings.—The mazzard cherry is a lofty,

rapid-growing, pyramidal-headed tree. Its fruit is small, dark brown, or black, with a sprightly flavor and slight bitterness. It is the original type of all the heart varieties.

Preparing and saving the Seeds.—The fruit is allowed to remain on the tree until thoroughly ripe. It is then shaken or picked off, and put into tubs, where the pulp is washed off until the stones are perfectly clean. They are then spread out on boards, and turned over occasionally until dry, when they are put away in boxes, mixed with sand very slightly moist. A layer of sand is spread in the bottom of the box, then a thin layer of the stones, next a layer of sand, and so on till the box is full. The boxes are secured against vermin, and put away in a cool, dry place, until needed for planting. If not planted in the fall, they may be wintered in a cellar, or out of doors, protected from rain by boards or other covering.

When to Plant.—If circumstances were favorable, all seeds would be better planted in the fall, or immediately after their maturity. Nature, in her course, indicates this to be a general law; but in cultivation this must depend on circumstances. The ground may not be in readiness. It may be so wet and heavy, that seeds would be so saturated with moisture during the winter as to lose their vitality; or the ground might become so beaten down and compact with fall, winter, and early spring rains, as to make it almost impossible for the young plants to make their way through it. All these things are to be considered in deciding the proper time to sow seeds. If the soil be very light and porous, cherry seeds may be sown as soon as gathered; if the contrary, it should be deferred till spring: but they germinate early and at a low temperature, so that it is necessary to keep them pretty dry and cool, and get them into the ground at the earliest practicable moment. We find it quite difficult to keep them properly, and yet prevent them from germinating before the ground is dry enough to receive them.

How to Plant.—For cherry seeds the ground should be *light*, in a good fertile state, but not strongly manured. The seeds are sown in drills as recommended for apple and pear seeds, and so thin as to give each plant space to grow in without being crowded by others. In this way, and with clean summer culture, the stocks will all be large enough at the end of the first season's growth, to be taken up and prepared for planting in nursery rows the following spring.

The *Mahaleb* (Cerasus mahaleb) is a small tree with glossy, deep green foliage. The fruit is black, about the size of a marrow-fat pea, and quite bitter. It blossoms and bears fruit when about three years old. It is considerably cultivated in many parts of Europe, as an ornamental lawn tree. There are very few bearing trees in this country yet; consequently nearly all the stocks used are imported, or grown from imported seeds.

The seeds are prepared, saved, sown, and managed in all respects similar to the mazzards, and are fit for transferring to the nursery rows at the end of the first season's growth.

The *common red pie cherry* and the *small morello* make very good stocks for dwarf trees of the duke and morello classes; but the hearts and Bigarreaus do not take on them. These are raised from seed in the same way as the mazzards and mahalebs. It may be added, however, as a warning, that buds are more liable to fail on them than on the mahaleb.

4th. *Stocks for the Peach.*—As a general thing the peach is worked on its own stocks in this country. The stones should be placed in a state of stratification during the winter, placed in boxes with alternate layers of sand or light earth, and be kept in a situation exposed to the frost; unless this is done they will not germinate the following spring; they require more moisture and exposure to open their hard shells, and induce germination, than any other fruit seeds. They should be examined a week or two before planting time, and if they exhibit no signs of vegetation more moisture should be given them; if they have been kept dry for a month or two before being stratified, they may require to be cracked. This is done by placing the edge of the stone on a wooden block and striking with a mallet; when cracked they may be mixed with moist earth and germinated in a warm place. The growth of every one so germinated can be depended on, and the rows will be regular. As the seeds are planted where the trees remain until transferred to the garden or orchard, it is a very good plan to nip off the point of the young root protruded from the seed; this makes it ramify, so that when taken up the trees have fine branched and fibrous roots instead of long tap roots, as is very generally the case.

Planting.—The seeds should be put into the ground as soon in the spring as it is in a fit state to be worked. A line is stretched, and holes made with a dibble to receive the seed; it should be put in with the root downwards, and be covered not over one fourth of an inch deep.

Plum Stocks are used for the peach in soils of a stiff, adhesive character, in which the peach does not succeed. In England the peach is worked almost exclusively on the plum, as it suits their moist climate and soil better. In France the hard shell almond is used almost exclusively on *dry*, and the plum on damp soils. Almond stocks are raised in the same way as the peach.

Dwarf Peach Trees are produced by working on the same stocks recommended for dwarfing the plum. Some time ago a French journal gave a very interesting account of experiments made in dwarfing the peach and plum, by a Dr. Bretonneau of Tours, France. He had succeeded in producing very pretty dwarf plums and peach trees on a dwarf plum indigenous to this country (*Prunus pumila.*) He exhibited beautiful prolific dwarf trees of the green gage plum on the sloe, and was making farther experiments with the dwarf almond as a stock for peaches. These subjects are all worthy of attention; we have many experiments of this kind under way, but it is yet too soon to communicate the results. The art of growing a large collection of fruits on a small spot of ground is of great importance to curious and tasteful people living in towns and villages.

Stocks for the Apricot and Nectarine.—Every thing that has been said of peach stocks, applies with equal force and propriety to these two trees.

5. *Stocks for the Plum.*—It is not a little difficult in this country to get good plum stocks. If seeds be taken promiscuously from any variety that is to be had, as is done with most other trees, the probability is, that of the seedlings not one in 500 will be suitable for a stock. I have seen bushels of seeds planted that were said to have been collected from strong growing trees, but out of the tens of thousands of seedlings produced from them, not 100 were ever worked, or fit to be. It is not only necessary to obtain seeds from vigorous growing trees, but from a species or variety that reproduces itself from seed. This is the point.

The *Horse Plum*, an oval, purple, free-stone sort, with vigorous downy shoots, reproduces itself from seed, and makes good stocks. On a suitable, well-prepared soil, its seedlings often attain two feet or more in height in one season, and are then fit for the nursery rows. They require a rich, substantial soil, prepared as recommended for pear seeds. Other vigorous sorts have been recommended in various parts of the country, but on trial they have been found quite inferior to the horse plum, and as a general thing worthless.

The *Canada* or *Wild Plum*, which abounds in Ohio, Michigan, and other western States, are distinct species, and reproduce themselves from seed. The seedlings of some grow extremely rapid, making fine stocks in one year on any good soil. They continue in a thrifty, growing state until late in the autumn ; but they should not be worked above the ground in the usual way, as their growth does not keep pace with the species to which most of our cultivated sorts belong. The best way to manage them is to take the yearling seedlings, whip-graft them on the collar, and set them out at once in the nursery rows ; they will make good trees for planting out in three years. The stock is all below the surface of the ground, and in time the graft sends out roots and becomes in a great measure independent of the stock. Where the seedlings are not large enough for grafting the first season, they may be set out in the nursery and allowed to grow one season, and then the earth can be removed from the collar until the graft be inserted, and then drawn up. To procure strong stocks for standard trees of weak growing sorts, like the *Green Gage*, such thrifty varieties as the *Imperial Gage* and *Smith's Orleans* may be grafted on this native species, and in two or three years they will make stocks strong enough for any purpose. The French use several natural species that are produced from seed—the *St. Julien*, large and small (Brussels of the English), and the *Damas noir*, large and small. The first is generally used for stocks for apricots and peaches as well as plums. We find none of these superior in vigor to the horse plum, but they are worked more successfully. In England, the *Brussels*, *Brompton*, and *Muscle* stocks are used, propagated from both seeds and layers. For *small sized garden trees*, either dwarf standards or pyramids, the cherry plum makes a very good stock. It is probably the same as used by the French under the names of "Cericette" and "Myrobalan." Several of our authors and even some English writers say that the *Mirabelle* is the stock used for dwarfing the *plum*, *peach* and *apricot*, but it seems probable that they are mistaken. In France the cericette or cherry plum is used, and stocks sent us from England as Mirabelle, are but the cherry.

How the mistake could be made is difficult to say, for the two trees are as different in habit, foliage, wood and fruit, as they can be. The cherry plum is a very low tree with bushy, erect branches, very straight, slender, willow-like, reddish shoots, exceedingly small leaves and buds, and smooth bark. The *Mirabelle* is also a low tree, but much more spreading than the other; the shoots are stouter, of a gray color and downy, with rather prominent buds for so small shoots. It ripens in September, and the cherry a month sooner.

The cherry plum is a natural species, and can therefore be produced true from seed. It maintains a vigorous growth all summer, and may be worked in July, August, or September. It may also be propagated from layers.

The *Sloe* is also used to some extent where very small trees are wanted, and we have no doubt some native species, as for instance the *Beach* and *Chicasaw* plums, small trees, will make good dwarf stocks. I am inclined to think, however, that very nice garden trees may be raised on the smaller species of the Canada Plum. The first year's growth and even the second are quite vigorous on them, but after that the vigor diminishes, and the trees become quite prolific. This and the cherry plum will probably become our principal stocks for dwarfing.

Plums for seeds should ripen well on the tree; they are then gathered, the pulp washed off, and the seeds dried and put away in boxes of sand in alternate layers, as recommended for cherries. They may be saved in fall or spring as circumstances already mentioned will admit.

Nearly all plums used for stocks may be propagated by layers. Mother plants or stools are planted out and cut back as recommended for paradise, &c.; the shoots of the previous season's growth are pegged down in the spring flat, and two inches of earth drawn over them. Every bud on these layers will produce a shoot that, generally, will be well enough rooted in the fall to be separated from the stool and planted out into nursery rows the following spring. These layered shoots are cut off close to the old plant, and the upright shoots produced during the previous season may be again pegged down.

The stools or mother plants managed in this way require the best treatment to maintain their vigor, that a supply of strong shoots may be produced every season fit to lay down in the spring. Weak, slender shoots, unfit to layer, should be cut out early in the season to aid the growth of those intended for use. This usually goes by the name of *Chinese* Layering.

Section 3.—Transplanting Stocks.

This comprehends three separate operations, *taking up*, *dressing* or *pruning*, and *replanting;* but before touching on the detail of these operations, it may be well to consider

1st. *The age at which Stocks should be transplanted.*— On this point there seems to be a diversity of opinion, not only among book writers but practical cultivators. The very general opinion, and one that is most acted upon, is, that they should remain where they have been propagated until they are large enough to be worked; a great many plans are therefore suggested for wintering seedlings, and especially the pear. The experience of the best cultivators every where is that seedling stocks especially, of all sorts, should be transplanted when *one year old*. It may be urged against this, that some seedlings are so small when one year old, as not to be worth transplanting; so feeble, that more care and culture would be required before they could be worked than they are worth. In reply, it can only be said that such feeble productions are only fit to be *thrown away*, because the seeds must have been defective, or the soil and culture bad; and stocks raised from poor seeds, or stunted by bad soil and culture, will never make sound, healthy, vigorous, or long lived trees.

When seedlings remain longer than one year in the seed bed, they grow up slender and weak; one more vigorous than its neighbors will ruin all around it, then the roots do not ramify, but continue to lengthen without forming laterals or fibres, and when removed and reduced to the necessary dimensions they receive a severe check; but at one year the check is very light, they at once form lateral roots, and instead of being drawn up tall and slender, they become stout and well proportioned. The best pear growers in Europe, and even in this country, would scarcely take as a gift two year seedling pears from the seed bed, unless in case of absolute necessity.

The proper plan is to take up all *seedling stocks*, and all *layers*, sufficiently rooted to bear separation from the stool, and all *cuttings* that stand close, *at one year old*,* and sort and arrange in separate classes, in this way: in one class put the strongest, those fit for immediate use, either to be grafted on the root, or budded the summer following; in another class, put such as may require to stand one year in the nursery rows to be fit for working; and in the third class, such as are too weak to be put in the nursery rows,

* The sloe (Prunus spinosa), or any such *very* slow growing thing excepted.

but will require to be "bedded out," that is, set closely in beds by themselves, where they can remain for one or two years, until they are large and strong enough for root grafting, or for the nursery rows. Unless in the case of stocks scarce and difficult to procure, this third class had better be thrown away at once, as it will cost as much to nurse them as to raise fine stocks from the seed.

2d. *Time to take up.*—There is but one proper time to take up all seedlings and rooted layers for stocks, and that is the fall, and this for several reasons. The first is, they are all liable to injury by the frosts of winter; seedlings have no side roots to hold them in the ground, and layers are near the surface, so that the freezing and thawing draws them up; the roots are thus exposed and seriously injured. The second is, they can be dressed during the winter in the cellar, and be ready for planting in spring. When taken up they can be laid closely in by the roots in the soil in a dry place, and covered over so as to exclude frost. When out-door work is over, they can be uncovered, taken into the cellar and dressed, and laid in again by the roots carefully in the same place, which should be protected from frost, of course, in the mean time. The third reason is, that when seedlings are taken up in the fall, the ground can be prepared for another crop; and this is of considerable importance. In the case of layers, the stools or mother plants can be manured, dressed, and put in order for another season's growth; and this, also, is important. Such are some of the advantages, or, in fact, the necessities of taking up stocks in the fall.

3d. *How to take up.*—Seedlings are very easily taken up, without in the least mutilating the roots, in two ways. If one person do the work, he should begin at one end of the row, and with a common spade, or, which is better, one with three strong prongs, a foot long and an inch and a half wide; dig under the plants without cutting the roots, and as fast as they are loosened below, pull them out, and in this way proceed. Another and quicker way is, for two men to loosen the plants, each on opposite sides of the row, inserting a forked spade as deep as the roots go, while another follows, and pulls out the plants. When the ground is quite soft, this way answers very well; but if dry or hard, the first is better.

Layers require more care and caution. A trench must be opened all around the layered branches deep enough to go quite below the roots, and in an oblique manner, so as to undermine them. Where the branches are pegged down, the pegs must be taken out, and the layer is then separated between the rooted part and the stool, and gently taken from the earth. Especial care must be taken not to split those that have been layered by incision; their removal must be done slowly and cautiously.

Mound Layers are easier separated; the earth is simply removed from the base of the rooted branches, and they are then separated within an inch or so of the stem.

Layered Branches or Chinese Layers.—When the young rooted plants are produced from the eyes of a buried shoot or branch, the pegs are removed, the whole branch dug under, completely loosened and separated from the stool; the young plants are then taken off one by one close to their base.

4th. *Pruning or Dressing Stocks.*—The objects in view always in performing this operation are, to remove injured or broken roots, to reduce the tap root that it may produce laterals, to reduce the stems to a proper proportion with the roots, and put them in a condition that will ensure a vigorous growth.

Seedlings taken from the seed bed, have always a long tap root, with few or no laterals; and as trees with such roots are unfit for safe transplantation, it is necessary to take measures to change their character. We, therefore, remove the small tapering portion of the root, as at fig 84, *A;* and this ensures the production of lateral or spreading roots near the surface of the ground. The pear roots especially are inclined more to descend in a straight line than to spread; and unless they are well cut back when young, they are always difficult to transplant safely afterwards. Roots that descend like the prongs of a fork, are usually destitute of fibres; whilst those that spread out horizontally, or near the surface, are well furnished with fibres, that not only make trees easily transplanted, but inclined to early fruitfulness. This operation on the roots, it is obvious, destroys the natural balance or proportion that existed between them and the tops. Hence the necessity for shortening the stem in a corresponding manner. But even if the roots were not shortened, the stems should be.

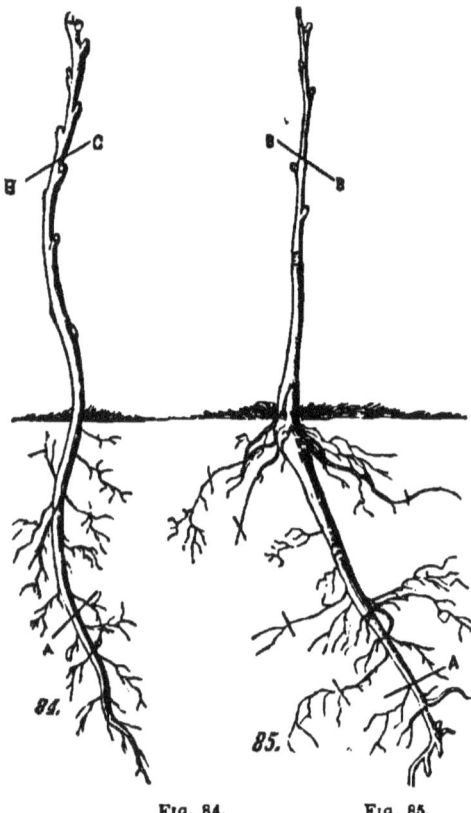

Fig. 84.   Fig. 85.

Fig. 84, a seedling stock, one year's growth, as it comes from the seed bed. The line at *A*, shows the shortening of the tap root. That at *B*, the shortening of the stem before replanting. *Fig.* 85, a quince cutting; the cross lines on the stem and roots indicate the pruning before replanting.

in order to obtain a vigorous growth. The very removal of the plant lessens the power of the roots to absorb and convey nutriment; and on this account, if no other, the stem should be reduced by way of regulating the supply and demand. We sometimes see young stocks planted out without any shortening of the stem; and the result is, they scarcely make any growth the first season—the roots are barely able to absorb enough to keep them alive. If one half the stem had been cut away, the remaining buds would have received such a supply of food as would have produced a vigorous growth. It is a pretty good rule, therefore, to reduce the stems of seedlings *one third* to *one-half*, as at *B C*, fig. 84; but there are exceptions to this. For instance, a stock with a very large and strong root, and a short, stout, close-jointed stem, well matured and furnished with plump, prominent buds, requires very little, if any shortening of the stem; and again, others are just the reverse, tall, slender, and feeble, having been suffocated in the seed bed. Such as these require to be shortened more than *half*, perhaps *two thirds*.

Layers or Cuttings (fig. 85) are in a different situation from seedlings, and require, therefore, different treatment. They have no tap roots, but masses of fibres; and these fibres, if they are preserved fresh and sound till replanted, need no shortening; but if destroyed by exposure, they should be cut off, to make way for new ones. The shortening of the stems depends entirely on the size and condition of the roots. If well rooted, and the roots be in good condition, they may be left a foot long; if poorly rooted they should be cut back to six or eight inches. This applies equally to the layers of the *quince, paradise, Doucain, plums,* etc.

5th. *Planting stocks in the nursery rows where they are to be budded.*—The first consideration which this operation suggests, is the *condition of the soil.* Under the head of soils, sufficient has been said respecting the modes of *deepening, draining,* and *enriching ;* and it is only necessary to say here, that where stocks are planted, the soil should be at once *deep, dry,* and *rich;* for no such thing as sound vigorous fruit trees can be raised on a poor, shallow, or wet soil. The various means of improvement have been already pointed out and explained. It may, however, be well to remark that ground may be *too rich,* and induce a rank, watery growth, that would either result in death at the final transplanting into the garden or orchard, or in a very feeble and sickly growth after it. We see frequent illustrations of this in the case of trees raised in old, worn out nurseries, where rapid growth has been *forced* by powerfully-stimulating manures. These rank, pithy, soft productions, are very attractive to the eye ; but they suffer so much by removal, no matter how well treated, that they seldom fail to disappoint the planter. This thing should, therefore, be guarded against. Manures used should be well decomposed, and incorporated with the soil, if possible the autumn before planting. A tree is not like a cabbage or a lettuce. The tenderness and succulency of these constitute their great merit; but the wood of a tree must be *firm, short-jointed,* and *mature,* and these requisites are always attained by a moderate and natural, not a *forced* growth.

Planting each species in the soil best adapted to it.— Where there are different characters of soils in a nursery, to be planted with a general assortment of stocks, it is important to give to each that which is best adapted to its nature; thus the pear, apple, and plum should have the richer, deeper, and more compact, or that with most clay. The plum in particular succeeds well on a pretty stiff clay. The cherry and peach should have the lightest and warmest

The *quince*, the *paradise*, and *Doucain*, do not require such a *deep* soil as the pear and the common apple seedlings, because their roots are fibrous and always remain near the surface; but it must not be inferred from t_is that a *shallow* soil suits these best.

6th. *When to Plant.*—In parts of the country where the winter is long and severe, or where freezing and thawing are frequent, fall planting cannot be successful, as the plants, having no hold of the ground, are drawn out and injured; and besides, if the ground is somewhat clayey and tenacious, the heavy rains that occur early in the spring will make it so compact that air will not penetrate it, and the young roots will form slowly and feebly. When neither of these difficulties is to be feared, fall planting is decidedly preferable. Spring planting should be done at the earliest moment the condition of the ground will admit, which is, when dry enough to crumble into fine particles when turned over with the spade.

7th. *Distance to Plant.*—We are all in the habit of planting quite too closely in the nursery; the consequence is that the trees are not well proportioned. As a general thing, the standards are in many cases as large six feet from the ground as at the collar, weak and top heavy, so that sticks have to be used to support them, even when four years old. Not long ago I observed in a nursery which has the reputation of being one of the best managed in this country, whole squares, some thousands of four year old apple trees, of all kinds, tied up to sticks; they were not able to support their own weight. One reason, and the principal one, was, they *were planted too close*, the other will be spoken of presently. *Pyramidal trees* are out of the question where such close planting is practised, the growth is always forced to the top. Nature gives us numerous and striking illustrations of the effect of close planting. We see in a natural group or thicket trees

6*

running up forty or fifty feet of an equal diameter, and without a branch; and if one such tree were left exposed, by the removal of those around it, the first high wind would blow it down. On the outskirts of this group or thicket, or perhaps completely isolated, in the centre of a field, we see another tree of the same species, branched almost from the ground, and with a diameter at the base twice as great as at half its height, and tapering upward with beautiful regularity, and capable of resisting a hurricane. To raise stout, well-proportioned trees, we must give them plenty of room, that they may have the advantage of air *all around*, and not only *at the top*.

There is scarcely a nursery to be found in which the trees are not grown too close—three or four on the space that one should occupy. There is to be sure great economy in close planting, for five hundred trees can be grown on the space that one should occupy, and with nearly as little labor; but it would really be better for people to pay twice or three times as much for their trees if grown so far apart that the air and light would have free access to them in all parts, and give them stout, well-proportioned forms. A reform in this respect is much needed, but it cannot be expected until purchasers become discriminating and intelligent on the subject.

The distance at which stocks should be planted in the nursery rows is governed entirely by circumstances. If it be intended to use a cultivator between the rows, they should not be less than *three and a half feet apart*. If spade and hoe culture be intended, two and a half to three feet will be sufficient. Where the trees are to be removed at the age of one year, one foot apart in the rows is sufficient; but if they are to remain until two, three or four years they should be eighteen inches to two feet. If removed at two years, eighteen inches is enough; but where standards remain three or four years, until they have heads formed, and pyramids remain until they have formed two or three tiers of lateral branches, two feet or two and a half is little enough. Indeed, when pyramids remain for three years, there should be a clear space three feet on all sides.

Dwarf standards require less space than full standards, and dwarf bushes still less. The stocks intended for these different classes of trees should be planted separately. In sorting the stocks at the time of dressing, the largest should be used for full standards and the smaller for low or dwarf standards.

8. *Mode of Planting.*—The square or plot of ground for each class of stocks being ready, a line is stretched along one side and a trench opened with the spade, deep and wide enough to hold the roots; the plant is then held against the side of the trench next the line, by one man, whilst the earth is filled in by another; when about half the earth is in, it is trodden down pretty firmly by the foot, and the remainder filled in. As buds are usually inserted on the north side of the stocks they should incline slightly to the south. Good pulverized surface soil should always be put upon the roots, to induce the immediate formation of young fibres. During the planting, the roots must be carefully guarded from exposure. A few only should be taken out of the ground at a time. When there are but few fibrous roots, puddling in thin mud is useful, otherwise not.

Planting Root Grafts.—The quickest mode of planting small root grafts is to stretch a line along the ground to be planted, and with a dibble make the holes and press the earth in around the plants. This dibble should be twelve to eighteen inches long, about two inches in diameter, pointed and shod with iron—fig. 86 represents one made of

Fig. 86
Form of Dibble used in planting root grafts.

the handle of a spade. One person will plant as many in this way as four could by opening trenches with spades. But where the plants are dibbled in, the ground must be in the best condition, perfectly dry and finely pulverized.

Treatment of Stocks after Planting.—The principal care which stocks require between the time they are planted and the time they are budded, is to keep the ground about them clean of weeds, and in a friable, porous condition on the surface by frequent stirring. The success of budding depends in a great measure on the condition of the stocks. They *must be* in a thrifty, growing state, and this can only be obtained with good treatment. Having now considered, in as much detail as seems necessary, the propagation and transplanting of *stocks* into the nursery rows, we proceed with

SECTION 4.—THE BUDDING, GRAFTING, AND MANAGEMENT OF TREES IN THE NURSERY.

The simplest and clearest method of treating this part of the subject seems to be, that of considering separately each year's operations in succession.

THE FIRST YEAR.—Strong yearling seedlings of the *apple, pear, cherry,* and *plum,* say one fourth of an inch and upwards in diameter, and well rooted layers of the *quince, paradise,* and *Doucain,* of the same size, planted in the spring in a good soil, and kept under good clean culture will, as a general thing, be in a fit state for budding in July, August, or September following. The budding may therefore be considered as the first season's work. The details of this operation may be divided for consideration, as follows:

1. *Time for Budding.*—2. *Preparation of the Stocks.*

—3. *Preparing the Buds.*—4. *Insertion of the Buds.*—5. *Untying.*

1st. The time for budding each species or class of fruits depends upon its habits of growth. Such as cease to grow early in the season, must be budded early, because it can only be done while the stocks are in a free, growing state, full of sap. Such as grow until late in the autumn, must be budded late, otherwise the new layers of wood formed after the insertion of the bud, would grow over and destroy it, or the bud would be forced into a premature growth towards autumn, which in fruit trees should always be avoided. The common sorts of plum terminate their growth early in the season, and are therefore budded early, whether with plums, peaches, or apricots, at Rochester usually about the last of July, or beginning of August. The native or *Canada* plum, and the *cherry* or *myrobalan*, grow freely till late in the fall, and may be budded in the latter end of August, or beginning of September. *Pears* on *pear stocks* are usually budded here in July, in anticipation of the leaf blight which stops their growth when it attacks them. Where no such thing as this is apprehended, they should not be budded before the middle of August, as the buds are not generally mature till that time. *Apples* on free stocks, and on the paradise and Doucain, may be budded as soon as the buds are mature, which is usually, here, about the first to the middle of August. *Cherries on free mazzard stocks*—as soon as buds are ripe here, about the first of August. *Pears on quince*, and *cherries on mahaleb*, not before the first of September, and from that to the middle of the month, as the quince and mahaleb grow late, and especially the latter. Peach stocks should always be budded the same season the seeds are planted, and, as they grow rapidly until very late, are not usually budded till about the middle of September. The budding period varies in different seasons. In a dry, warm season, the young wood matures earlier, and stocks cease to grow sooner, and are, therefore, budded earlier than in a cool, moist season, that prolongs the growth of the stocks, and retards the maturity of the buds. Stocks growing feebly require to be budded earlier than those growing freely. It is necessary to keep an eye to all these points.

The destruction of insects must be promptly attended to. An army of *slugs* may devour the foliage of the pear and cherry, and even the plum, in a day or two, and prevent their being worked that season. The *aphis*, too, frequently appears in such multitudes as to check the growth. Dry lime or ashes thrown on the slugs will kill them, and strong soap suds, or tobacco water, so strong as to assume the color of strong beer, will kill the aphis.

2d. *Preparation of the Stocks.*—This consists in removing such lateral shoots from the stock as may be likely to obstruct the insertion of the bud. Our practice is to do this at the moment of budding, one person doing the work in advance of the budders. If done a few days previous, and several shoots are removed, it checks the growth of the stocks, and they do not work so well. It might answer very well to do it two or three weeks previous, so that they might recover from the check before being budded.

3d. *Insertion of the Bud.*—Having treated so fully of the manner of preparing and inserting the buds in the article on budding, nothing farther need be said on these points here.

In free stocks the bud should be inserted within three or four inches of the ground.

In some parts of the west, Wisconsin, Illinois, and some other places, certain rapid, late-growing, and rather tender varieties are liable to be winter-killed if budded close to the ground, probably by the sudden thawing of that part caused by the refraction of heat from the ground. In view of such a difficulty, it may be well enough to bud high up, but, as a general thing, low budding makes the best trees All dwarf stocks should be budded as close to the surface of the ground as it is possible, and even some of the earth may be removed and put back when the budding is done. The necessity for this lies in the fact that all dwarf stocks should be wholly below the ground when finally planted out in the garden or orchard.

4th. *Untying the Buds.*—In ten days or a fortnight after the buds are inserted, they should be examined, and such as have failed may be budded again if the stocks continue to grow. In some cases it may be necessary, and particularly with cherries, to loosen the buds and tie them over again, as rapid growth will cause the string to cut the bark before the bud has completely united, or is fit to be untied. This seldom occurs, however; as a general thing, the strings may be removed in three weeks to a month after the budding; and they should never be left on over the winter, as moisture lodges around them to the detriment of the bud. As soon as the budding is done, the ground should be worked over with the cultivator or forked spade. The first season's management of stocks too small for budding consists simply in keeping the soil clean and mellow, and in guarding against the attacks of insects.

The treatment of *root grafts* the first season consists in cleaning and loosening the ground, the removal of suckers from the roots as fast as they appear, and pinching early any strong side shoots likely to weaken the leader.

SECOND YEAR.—Where the buds failed the previous season, the stocks should now be whip-grafted near the surface of the ground. They will be little behind the buds, and will make nearly as good trees, if neatly done. Plums and cherries must be done before, or as soon as the buds begin to swell (say in March here); pears and apples may be done later. The *second sized stocks, planted last season*, and intended to be budded this, should, if in a feebly growing or stunted condition, be cut back to within two or three inches of the surface of the ground. This will give the roots new vigor, and thrifty shoots will be made by budding time that will work more easily and successfully than the old stock. In a month or so after being cut down, all the shoots but the strongest one should be removed. The *stocks budded last season* are headed down to within three or four inches of the bud, just as the leaves are beginning to appear, and all buds starting into growth on the stock, either below or above them, rubbed off.

Treatment of the growing bud consists in keeping all shoots that appear on the stock rubbed off. If side shoots appear early, and are likely to contract the growth of the leader, they should be pinched off. Any that assume a reclining or crooked habit should be tied up to the stock, or to a support, which may be a wooden pole four feet long, sunk a foot in the ground at the root of the stock; both the stock and growing shoot should be fastened to it (fig. 87), but not so close as to impede the growth. This is only necessary with certain weak, irregular growing sorts. In August the portion of the stock left above the bud at the heading down in the spring should be removed with a sloping cut, close and smooth, as at *A* (fig. 87), at the highest point of union between the bud and stock. The new layers of wood made after this time covers the wound before growth ceases in the fall. Side shoots, when they appear, must

Fig. 87.
A young budded tree in its first season's growth, supported by a stake. The line at *A*, indicates the cutting away of the stock close to the bud.

be checked, if too vigorous, by pinching off their ends, but not entirely removed, as they assist in giving size and strength to the lower part of the body of the young tree. The peach almost invariably produces numerous side branches the first season, and it is a very common but very erroneous practice to prune these all off in mid-summer. The proper course is to maintain an uniform vigor amongst them by pinching, and to prevent any from encroaching on the leading shoot; in this way we get stout, well-proportioned trees. This brings us to the end of the second year, and gives us young trees of one year's growth. *Peach trees* should always be planted out at this age, and all trees intended for training in particular forms; but as this part of the subject will be considered under the head of "*Selections of Trees*," we will proceed to the course of management for the

THIRD YEAR.—We commence this year with trees of one year's growth; and the first point is to determine what *form* is to be given them, whether *tall* or *dwarf standards*, *pyramids*, *bushes*, or *espaliers*. Having settled these matters, we have but to follow up the proper course to accomplish the desired ends. It may be well to take each of these forms in succession, and point out the necessary management under various circumstances.

1st. *Standards.*—Until very lately, trees of all sorts, and for every situation, were grown as tall standards, with naked trunks six and even *eight feet* high. Indeed, it appeared as though an impression existed amongst people that a tree was not in reality a tree, nor worthy of a place on their grounds, if it had not this particular form. Latterly, however, since fruit tree culture has become more practised, and somewhat better understood, this impression has been gradually losing ground, and in all parts of the country low trees are finding advocates.

Experience is beginning to teach people that whilst tall

standards in an orchard possess the single advantage of admitting the operations of the plough under the branches, low standards are much more secure against the numerous fatal diseases that attack the trunks—are much more accessible for the performance of all the necessary details of management, and for the gathering of the fruit.

These are all very important advantages certainly; but the most important one is the safety of the tree against diseases of the trunk. In all parts of this country, we have a powerful sun in summer, and in winter and spring sudden and violent changes from one extreme to another; and experience has shown, that the trunk and large branches, being fully exposed to all external influences, are generally the parts first attacked with disease. Cultivators are, of course, at liberty to choose for themselves; but, except to meet the wants of some particular circumstances, no standard tree should have a branchless stem above *five* feet in height: *four* is preferable for all, except orchards of common apples for cider or stock. Trees with heads only four feet from the ground, are always easy of access, and the natural spread of the branches affords a great protection to the trunk at all seasons. Nurserymen should by all means encourage by precept and example the cultivation of low-headed trees.

Starting with the yearling trees for standards, we examine the habit of the variety, whether stout or slender, whether branched, as many varieties are the first season, or without branches. Before proceeding to the operation of cutting down to increase the size of the trunk, the reader is referred to the principles and practices of pruning in the first part of the work. No pruning should be attempted for the attainment of any special purpose without having first carefully studied these.

If slender and without side branches, as in fig. 88, they should be cut back twelve to twenty inches, as at *A*

BUDDING, GRAFTING, ETC. 139

Fig. 88.   Fig. 89.

Fig. 89. a y'rling tree; from the bud A, indicate the cutting back to make a stout *n for a standard. B and C. the cutting back for pyramids or low standards. D. the cutting back for dwarfs or espaliers.
Fig. 89. a yearly tree once cut back to form trunk for a standard.

this removes the buds that would push first, and retains the sap in the lower parts, which will give a stout body. The taller and more slender the tree, and the smaller the buds, the farther it becomes necessary to cut back. In fact, some very feeble growing sorts must be cut back till within a foot or less of the base. During the summer, trees cut back in this way may produce lateral shoots on the greater part of their length. These must not be pruned off, but kept in an uniform size and vigor, by pinching any that threaten to exceed their proper bounds. The shoots immediately below the leader, must be watched, as they are always inclined to push too strongly.

A tree thus cut back, and the side branches regulated by pinching, will, in the fall, have a stout body, and present the appearance of fig. 89. Where the yearlings are short and stout, and are furnished with a few lateral shoots, cutting back may be unnecessary. The largest of the side shoots may be pruned off wholly, and the small ones left to retain the sap in the lower part of the stem, at least till midsummer, when new ones will have been produced. There are certain stout-growing, branching varieties of all the fruits that require no shortening and very little pruning of any kind, to form stout trunks, and especially when not planted too close.

Dwarf Standards.—The management of yearly buds to produce these, is similar to that described for standards, varying it always to suit the particular habit of the species or variety; tall slender growing sorts require cutting back, and the suppression of branches at the top; but many varieties of cherries and plums, some very stout growing pears and apples, and all apricots and peaches, may commence the formation of heads this season. The stem is cut at the point desired, two to three feet from the ground, to form the head on, and three or four of the stoutest shoots, growing in opposite directions, are preserved, whilst all others close to them are pinched off, when two or three inches long; side branches are allowed to remain that season on the stem to strengthen it, but they are kept short and regular by pinching. In the fall these trees will be fit for the final planting out, whilst those of weaker habit will require another season, if they be wanted with heads.

Pyramids.—Yearling trees intended for pyramids are cut back so far as to ensure the production of vigorous side branches within six or eight inches of the stock. The habits of growth of the species and variety must be carefully taken into account. Some are disposed, from the beginning, to form lateral branches, and others require vigorous measures to force them to do so. As examples, the Bloodgood pear is very much inclined to branch the first year, whilst the Louise Bonne de Jersey and Duchess d'Angouleme seldom do so, unless in some way the growing point be checked. So it is in cherries; most of the Dukes and Morellos are inclined to produce laterals the first season, but the free growing sorts, *Hearts* and *Bigarreaus*, rarely do so, unless the point is checked early in the season. So it is in all the fruits, and therefore no general rule can be given, but the appearance of the tree indicates the treatment required. Where we see side branches *naturally* produced the first season, we at once conclude that the buds are well disposed to break, and the cutting back may be comparatively light. Where no side branches are produced, we must be governed by the appearance of the buds on the lower part of the tree, where it is desired to produce the lower branches; if they be small and flat, it will take close cutting to arouse them, but if plump and prominent, less vigorous measures will be necessary. In the case of short, stout, and branched yearlings, a few of the best placed, lowest, and strongest branches are reserved, whilst the others are entirely removed. We then shorten the reserved branches according to their position, leaving the lowest the longest. The leading shoot is shortened, so that all the buds left will be sure to push and form shoots. When these have attained the length of two or three inches, the strongest and best placed are selected for permanent branches, and the others are pinched off.

Yearlings that have no side branches, figure 88, we generally cut back one half as to *B*, and in many cases two thirds to *C*, in order to obtain strong branches near the ground. Every bud below the one we cut to, should push, and when shoots of two inches or so are made, we select two, three, or such number as may be wanted, of the strongest and best situated to be reserved, and pinch the others. It very generally happens that two or three buds next below the one we cut to, push with such vigor as to injure both the leading shoot above and the side shoots below them. They must be watched and pinched as soon as this disposition becomes obvious. Yearling trees managed in this way will present in the fall the appearance of fig. 90.

Purchasers are very apt to favor *tall* trees, even at the expense of their forms; and nurserymen, even those who know better, with a view to suiting the tastes of their customers, rarely cut their trees back sufficiently to make pyramids. The first branches are seldom less than two feet from the ground, and it is quite difficult to make nice pyramids of such trees afterwards; at all events, it incurs a great loss of time, for the whole of the branches and half of the stem must be cut away to produce the required form.

Dwarf Bushes.—The apple on paradise is generally grown in this form, with six to twelve inches of a stem and spreading heads. The Morello cherry and the cherry and Mirabelle plums, and many kinds of pears, may be grown as dwarf bushes, if desirable. The stocks must all be of a dwarf character. Plants from which the strongest have been selected for dwarf standards and pyramids, will make very good bushes. The branches being so near the root renders a less amount of vigor necessary. Very strong yearling plants may be allowed to form heads the second year, but such as are *very* slender will require cutting back and another season's growth, before the head is allowed to form; and they will require a similar course of treatment, as has been recommended for standards, and dwarf standards. No matter what the character of the tree is, a *stout stem* is necessary, and although the measures taken to obtain this seem to require in some cases a loss of time, still there is a gain in the end; for trees allowed to form heads before the stems are amply sufficient to support them, require a great deal of extra care after planting out, and a course of shortening back, that

Fig. 90.

Fig. 90, a two year old tree cut back once, and intended for a pyramid. The cross lines indicate the second cutting back.

offsets the temporary advantage of forming the head a year sooner. This holds good in all cases. The mode of forming the heads of dwarf bushes is similar to that described for standards.

Espalier Trees.—These have a few advantages peculiar to themselves, which will be explained under the head of " the selection of trees for the garden."

To form espaliers, yearling trees are usually chosen, planted in the place where they are to remain and cut back to within four or five buds of the stocks, as at D, fig. 88; these buds break and produce shoots from which the strongest are chosen to form the arms, and the others are rubbed off.

The peach grows so vigorously that, if the growing bud be checked when a foot high, it will produce side shoots, from which two may be selected from the main branches of the espalier, and thus a year will be saved. Another way is to insert two buds, one on each side of the stock. Very nice espalier trees may be grown in the form of a pyramid with a main stem and lateral branches, the lowest being the longest. Trees for this form require the same management as pyramids, except that the branches should be placed opposite on *two* sides. This brings us to the end of the third year, and the trees are now two years old from the bud. At this age we take it for granted that all trees on dwarf stocks for *pyramids*, *dwarfs*, and *espaliers*, and all standards even, of the peach, apricot, and nectarine, and in most cases the cherry and plum, will be finally planted out. Standard pears and apples are almost the only trees that require to be left longer in the nursery, and their management during the third and fourth years of their growth, if allowed to remain so long, will be similar to that described for the second. In the spring, February or March, the leading shoot is cut back in order to increase the stoutness of the stem as it advances in height; and during the summer, the side shoots are kept of uniform length and vigor by pinching. The lower side branches are removed gradually every season as the tree becomes strong enough to dispense with them. As it has been before remarked, the cutting back depends always on the natural character of the subject—stout, short-jointed, moderate growing sorts, that *naturally* increase in height and diameter of stem in proper proportions, will require no cutting back. Very few, however, have this habit. In nearly all cases more or less shortening in, every spring, is necessary until the stem has arrived at the requisite height, and is well proportioned, decreasing gradually in diameter from the base to the top.

The Treatment of the Soil.—During the whole period the trees remain in the nursery, the ground about them must be kept clean and finely pulverized on the surface by repeated and continual stirring. Every spring, as soon as the heavy rains are over, and the ground settled and dry, the space between the rows should be ploughed, if they are far enough apart to admit of it. A small one-horse plough, such as is used for ploughing cornfields (see implements), is suitable, but it should not be allowed to go nearer the tree than six inches, nor so deep as to come in contact with the roots. After ploughing, the cultivator may be run through once each way between the rows, every week or two, and this will leave very little hoeing to be done. If the rows are so close as not to admit the plough and cultivator, the forked spade must be used in the spring to give the ground a thorough stirring, and the hoe afterwards. If the ground be naturally adhesive, a second or even a third ploughing or spading may be necessary in the course of the summer; for it must at all times be kept in a loose, porous condition, or the roots will be deprived of the benefits of the air and moisture. Stirring the ground so often that weeds barely make their appearance, is not only the best, but most economical culture.

It need scarcely be added that in using the plough or cultivator among trees, a very short whiffletree should be used, the horse should be gentle and steady, and the ploughman both careful and skilful; and laborers who use the spade or hoe, should be duly cautioned against cutting or bruising the trees with their implements.

SECTION 5.—PROPAGATION AND NURSERY CULTURE OF SEVERAL FRUIT TREES AND SHRUBS NOT USUALLY GRAFTED OR BUDDED.

1st. *The Grape Vine.*—This is one of the easiest subjects to propagate among all our fruit trees.

In all stages of its growth it should have a *dry and rich soil, dryness* first and most of all. The surest method of propagation for unpractised hands, is *layering.* A branch or shoot of the current season's growth, laid down in June, in the manner described in the first part of this book (figs. 61 and 62), will be well enough rooted to bear transplanting in the fall or spring following. The reader is referred to the instructions on layering.

The next mode is by *long cuttings.* At the winter pruning, the strongest, roundest, and firmest shoots of the previous season's growth are selected, and cut into pieces twelve to eighteen inches long, with two or three eyes, as in fig. 60. They are cut close to an eye at the lower end, or a piece of the old wood may be attached, like fig. 58. These cuttings are buried in dry, sandy earth, till the ground is fit to receive them in the spring.

In planting, the whole cutting is buried but one eye, and some cover that even as much as an inch deep. The long cutting must be laid in the trench obliquely, as in fig. 60, so that the lower part will not be out of reach of air and heat, without which new roots will not be formed.

During the summer, the earth must be kept clear and friable around them; and, in dry seasons, a thick mulching will be very beneficial in preserving a uniformity of heat and moisture. In the fall, the plants will be fit for final transplanting; but if they remain another season, they should be pruned back in winter to two or three buds at the base, and during the following summer only one or two shoots be allowed to grow, all others being rubbed off early.

Layers, when taken from the mother plant, and set in nursery rows, should be cut back in the same manner, in order to obtain one or two vigorous shoots when the plant is to be finally set out.

Short Cuttings.—These consist of only one eye, from the stoutest and firmest shoots of the previous year's wood (fig. 59), with not more than an inch of wood on each side of it. These cuttings, however, seldom succeed so well in the open ground as others. They require a little artificial bottom heat.

The simplest way to treat them is to make a sort of hotbed, with two to three feet of half-decayed stable manure, well mixed, and six or eight inches of light sandy soil. The cuttings are planted in this a quarter to half an inch deep, and covered with a glazed sash. If carefully and regularly watered, and well ventilated, they will make fine plants by the autumn. A better way than this is, especially in propagating the foreign varieties, to put them into pots, and put the pots in the hotbed.

A single cutting may be put into a small three inch pot, covered a fourth of an inch deep; or several cuttings may be inserted in a larger pot. In this case they should be placed around the sides. When they have made a growth of about six inches, they may be shifted into larger pots, with good, rich compost. In one season they will make good, strong plants. Plenty of air should always be given them, as soon as they are rooted, to prevent their being drawn up into weak, watery shoots. When the native hardy sorts are raised from eyes in the hotbed as described, the yearling plants should be pruned to a couple of eyes, and transplanted into nursery rows where one season's growth will fit them for final setting.

Single eyes, in all cases, make the best plants.

2d. *The Currant.*—Every one knows how to propagate this. A yearling shoot, six inches to a foot long, taken off close to the old wood, and planted half or two thirds its length in the ground, in the spring, will make a strong, well-rooted plant in the autumn. To prevent shoots from springing up below the surface of the ground, the eyes on that part are cut out, or they may be left the first season, and cut out when the plants are rooted.

The buds aid in the formation of roots. When a variety is rare and scarce, the young shoots may all be layered in July, and they will make well-rooted plants in the fall.

3d. *Gooseberries* are propagated in the same way, and with almost equal facility, as currants, though, as a general thing, they do not grow with such rapidity. Layers are the surest, but they require to be one year in the nursery rows after being separated from the mother plant to make them strong enough for the final planting. An inch or two of swamp moss laid over the surface of the ground in which layers are made, assists in retaining the moisture. This is applicable to all kinds of layers.

4th. *Strawberries* are propagated by the runners, which spread on the surface of the ground in all directions from the plant as soon as it begins to grow in the spring. Where a variety is scarce, and it is desirable to multiply it carefully, these runners should be sunk slightly in the ground, and pegged down with small hooked sticks, as they will root and form plants fit for removal much quicker than if left to root in their own way. With good management, a single plant may produce twenty-five to fifty, and even one hundred in one season. Plants to be propagated from, should have abundance of space, and a deep, rich soil. An application of liquid manure will stimulate their vigor, and increase the number and strength of the runners.

5th. *Raspberries* are propagated from suckers, or shoots produced from the collar, or spreading roots of the plant. They are renewed every season. The canes bearing but once, they may be propagated by layering the young canes in midsummer, and by cuttings of the roots. The latter mode is advantageously applied in the case of new or rare sorts.

6th. *Berberries* are propagated by seeds, suckers, and layers, in the simplest manner. Rare sorts are also grafted successfully on the common ones early in the spring, in the cleft mode.

7th. *Mulberries.*—The large black mulberry is the only one worthy of culture for the fruit. It is easily propagated both by cuttings and layers. The latter mode is the surest.

8th. *Chestnuts.*—The common American chestnut may be propagated from seeds either planted in the fall or kept in sand all winter, and planted early in the spring. In one season they are fit to transplant into nursery rows, and in two years more at most may be finally planted out.

The Spanish chestnut is propagated either from seeds or by grafting on the common chestnut. Its fruit is three times as large as the common.

9th. *Filberts* are propagated either from suckers or by grafting. If seedlings are used for stocks, the grafted plants are the best, as they are not only more prolific, but they do not throw up suckers. They may be grown either as low standards, with stems three feet high, or as pyramids or dwarf bushes.

10th. *Walnuts* are propagated from seeds or by grafting, in the same way as filberts. There is a *dwarf prolific* variety, that bears quite young, and makes handsome pyramidal garden trees.

SECTION 6.—LABELS FOR NURSERY TREES.

It is highly important that a correct system for preserving the names of varieties be adopted. Our practice is, to make labels of cedar, eighteen inches long, three inches wide, and about an inch thick. These are pointed on one end, to be sunk in the ground eight or ten inches, and the face is painted white. When a variety is to be budded or grafted, the name, or a number referring to a regular record is written on it, and it is put in the ground in front of the first tree of the variety. Besides this, we invariably record in the nursery book each row, with the kind or kinds worked on it, in the order they stand in the square. In case of the accidental loss of the labels, the record preserves the names. Figure 91 represents this kind of label, and though there are many others in use, we believe this is one of the simplest and best.

At the time of budding or grafting, we usually write the name on with pencil, and after the square has been all worked, the numbers are made with a brush and black paint.

Fig. 91, label for nursery rows.

Section 7.—Taking up Trees from the Nursery.

This is an operation that should be well understood, and performed with the greatest care. The importance of the fibrous roots has been already explained. It has been shown that they are the principal absorbing parts of the roots, and when they are destroyed the tree receives a great shock, from which it requires good treatment and a long time to recover. There is a great difference in the character of roots, some penetrating the ground to a great depth, and requiring much labor in the removal, others quite fibrous near the surface, and consequently very easily taken up. This difference is not owing alone to the difference in the species, but to whether the subjects have or have not been frequently transplanted. The way to take up a tree properly, is to dig a trench on each side at the extremities of the lateral or spreading roots, taking care that the edge, and not the face of the spade, be kept next the tree, so that the roots will not be cut off. When this trench is so deep as to be below all the lateral roots, a slight pull, and a pry on each side with the spade, will generally bring out the trees. If there be strong tap roots, running down to a great depth, they may be cut with a stroke of the spade. Laborers who have not been accustomed to the work, invariably perform it badly, and it is difficult to get it properly done even by experienced hands. It is a work requiring care and leisure, though it is usually performed slovenly and in great haste.

Labelling.—When a tree, or a number of trees, of any variety are taken up, a label, with the name written on it, should at once be attached. The kind of label used in the nurseries here, is a piece of pine about three and a half inches long, three fourths of an inch wide, and one eighth of an inch thick. A neck is made on one end by cutting into each edge about an eighth of an inch; a piece of No. 32 copper wire, about seven or eight inches long, is then fastened in the middle, on the neck of the label, with two or three twists. The two ends of the wire are then placed around the stem, or a branch of the tree, and are fastened with a twist or two. This kind of wire and label we find by experience to be not only safe, but more expeditiously attached than any other. If a little paint is rubbed on just before being used, the writing will be more legible and permanent, but it should be so light as to be barely perceptible, else it will clog the pencil. These labels are made very quickly, as follows: take a common inch board planed, cut into pieces the length of the label, make a groove with a knife or saw along both sides, at one end for the neck, and then set the piece on its end, and split off the labels with a knife; this can be done nearly as fast as one person can pick them up. The wire costs three shillings per pound, and is cut into lengths with a pair of common shears.

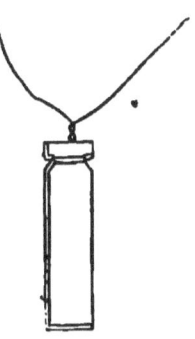

FIG. 92.

Wire label for trees.

Packing.—Persons who are ignorant of the structure of trees, never appreciate the importance of packing; and that is the reason why so many trees are every year destroyed by exposure. It is not uncommon, in this part of the country, to see apple trees loaded on hayracks, like so much brush, without a particle of covering on any part of them, to travel a journey of three or four weeks in this condition. Of course it is utterly impossible that such trees can live or thrive; and yet the persons who thus conduct their nursery operations, are doing the most profitable business. Such practices are not only dishonest, but highly injurious and disreputable to the trade; and it is by no means fair to class such people amongst respectable and honorable nurserymen.

Purchasers are often at fault in this matter. Nurserymen have to buy and pay for the material used in packing. Mats cost one to two shillings apiece; straw, three cents per small bundle; yarn, one to two shillings per pound; moss, three to four dollars per load, in many cases; and besides, the labor of packing, when *well done*, is very great. It is, therefore, not unreasonable that a charge be made; but some people, rather than pay twenty-five or fifty cents for packing fifty trees, would expose themselves to the risk of losing all. Purchasers should invariably charge the nurseryman to whom they send their orders, to *pack in the best manner*. Better pay one or even two cents per tree for packing, than lose it or injure it so much as to make it almost worthless.

The mode of packing pursued here is this: Where the trees are packed in bundles, a number of ties are first laid down, then a layer of long rye straw, three or four inches deep; the trees are then laid compactly together, straw being placed among the tops to prevent their being chafed when drawn together, and damp moss from the swamp is shaken among the roots. When the bundle is built, long straw is placed on the top as below, and it is then bound up as tightly as it can be drawn. Straw is then placed around the roots sufficiently thick to exclude the air, and then a bass mat is sewed on over the straw. If the bundle is only to go a short distance, the straw can be so secured around the roots that the mats may be dispensed with; but if it has a long journey to perform, it should be matted from bottom to top, and sewed with strong tarred spun yarn, about as thick as a goose quill. *Boxes* are rather more secure for very long journeys; they should be made of white wood, or some light timber that holds nails well. If the trees are composed of several varieties, they should be tied in small parcels of four to six each, according to the size. The sides and ends of the box should be well lined with straw, and the roots bedded in moss and the tops in straw, to prevent chafing.

If the box be large, two rows of cleats are necessary—one in the middle and one in the top, to hold the trees in their place and to keep the box from spreading. When the box is nailed up, it should be banded at both ends with iron hoops, fastened with wrought nails. Packed in this way, trees may go any distance with safety. The season of the year modifies the mode of packing. The roots should always for a long journey be immersed in a thin mud before being packed, as this excludes the air; but in the fall, this mud should be dry before the package is made up, and the moss should contain very little moisture. In a frosty time the less moisture there is about the roots the better; but an abundance of straw should be used to exclude the air and frost.

Heeling in.—When trees are taken up, and cannot be either packed or planted at once, they are laid in by the roots in trenches; the longer they have to remain in this situation the better it should be performed. Trees are often wintered in this way, and if the trenches are dug deep, and the roots well spread out and deeply covered, they are perfectly safe. It should be done in such cases with almost as much care as the final planting of a tree. When great bundles of the roots are huddled in together, and only three or four inches of earth thrown over them, both air and frost act upon them, and they sustain serious injury. Tender trees likely to suffer from the freezing of the shoots, should be laid in an inclined, almost horizontal position, and be covered with brush, evergreen boughs, or something that will break off the violence of the wind and frost. Straw should not be used, as it attracts vermin. Some rough litter or manure should also be thrown around the roots, and in this way the most tender of all our fruit trees may be wintered with safety.

PART III.

THE LAYING OUT, ARRANGEMENT AND GENERAL MANAGEMENT OF DIFFERENT KINDS OF PERMANENT PLANTATIONS OF FRUIT TREES, SELECTION OF TREES, AND VARIETIES, AND PRUNING AND CONDUCTING TREES UNDER VARIOUS FORMS.

CHAPTER I.

PERMANENT PLANTATIONS OF FRUIT TREES.

SECTION 1.—THE DIFFERENT KINDS OF PLANTATIONS.

THESE are of several kinds, and may be classed as follows—1st. The *Family Orchard*, which is a portion of the farm set apart for the production of the more hardy and common fruit, principally apples, for the use of the farm stock and the family. 2d. The Market or Commercial Orchard, is a large plantation of the various species of fruit trees for the production of fruit as an article of commerce. 3d. The *Fruit Garden*, with the Farmer is a plot of ground near the dwelling, in which the finer fruits, as pears, peaches, plums, cherries, apricots, &c., and all the small fruits are cultivated. In many cases, and even in most cases, it is a portion of the kitchen garden, where the table or culinary vegetables are grown. With the professional man, the merchant, the mechanic, and others who reside in cities, villages, and their suburbs, possessing but small tracts of land, at most but a few acres, the fruit garden is the only source for the supply of fruits for their families, and is usually planted with the more rare, perishable, and valuable sorts that cannot so easily be procured in market.

The pleasure and profit derived from fruit plantations,

under any or all of these circumstances, depend upon *the judicious selection of soil, situation, trees, and varieties,* and *their proper arrangement and management.* These are the essential points, and every man who contemplates planting to a greater or less extent, should avail himself of all the light which experience has shed upon these various branches of the subject, before making the first movement towards the execution of his project.

Section 2.—The Orchard.

The orchard is distinguished from the fruit garden in this, that the trees planted in it are generally of the largest size to which the species attain; they are grown in the natural, or, as it is called, standard form, without any particular training, and the varieties are generally the most hardy and productive of the species.

1. *The situation of an orchard* with regard to exposure or aspect, requires very little consideration in some parts of the country. Where, as in Western New York for instance, the winters are uniform, or comparatively so, in temperature, and late spring frosts do not prevail, the main difficulties to guard against are the prevailing high winds from the west and north that injure the blossoms and blow off the fruit before it is mature. If possible, a situation should be chosen where some natural obstacle, as a hill or a belt of woods, would break the force and influence of these destructive winds. Where no such obstacle naturally exists, a belt or border of rapid growing trees, such as *soft maples, white pines,* and Abeles, should be planted simultaneously with the planting of the orchard, that they may grow up and form a protection by the time the trees have come into bearing.

In other sections, as in some of the central and southern counties of New York, and in some parts of Ohio,

Illinois, Wisconsin, and others of the western as well as in the southern States, where late and fatal spring frosts prevail, the selection of a situation is a most important point. In such localities an eastern and southern exposure, and low grounds, are to be avoided.

John J. Thomas, in his Fruit Culturist, states that, " In the valley of the Conhocton, which is flanked by hills five hundred feet high, peach trees have been completely killed to the ground, but on one of the neighboring hills, five hundred feet above, and probably twelve hundred feet above the level of the sea, an orchard planted in good soil yields regular crops. In the town of Spencer, Tioga County, near the head of Cayuga inlet, peaches have withstood the climate and done well at an elevation of seven hundred feet above Cayuga Lake." Lawrence Young, Esq., Chairman of the Kentucky Fruit Committee, reported to the Pomological Convention at Cincinnati, in 1850, the case of an orchard in that State, lying within the peach district, occupying the slopes of hills of no great height, inclining gently toward a river distant only a few hundred yards. Its success was that common to a·fickle western climate—a fruit year and a failure, or perhaps two years of productiveness and three of disappointment in every five.

Within five miles of this orchard, however, is located a hill six hundred feet high, upon which the peach crop has not failed since he first knew it. Numerous other instances are quoted and the particulars given with great accuracy, showing the effects of even very slight elevations.

Among others is an instance of the heath-peach bearing a full crop in one part of an orchard, whilst in another part thirty feet lower, the same variety bore not a single fruit. Multitudes of such cases might be collected in all parts of the country where the climate is variable, because in such situations vegetation is earlier excited than in those more elevated and colder, and frosts always fall more heavily on low than on high grounds. Every one who has paid the slightest attention to the action of frost on vegetation is aware, that even an elevation of two or three feet of one portion of the same field or garden above the other frequently proves a protection from an untimely frost. In a dry and firm soil, vegetation is more exempt from injuries by frost than in a damp, soft, and spongy soil on the same level, not only because trees on such soils are more mature and hardier in their parts, but because the soil and the atmosphere above it are less charged with watery particles that attract the frost. Bodies of water that do not freeze in winter, such as some of our inland lakes, exert a favorable influence for a considerable distance from their margins in protecting vegetation from late spring and early autumn frosts.

In some parts of the West, as in Wisconsin and Illinois, the winters are so variable—during the day as mild as spring, and in the night the mercury falling many degrees below zero—that even apple and pear trees in soft, damp, and rich soils, are frequently killed to the ground.

In such localities, experience has taught cultivators that elevated, dry, firm, and moderately rich soil, that will produce a firm, well-matured growth, is the only safeguard against the destruction of plantations in the winter. In all localities where fruit culture has made any considerable progress, there is generally experience enough to be found, if carefully sought for and collected, to guide beginners in fixing upon sites for orchards; and no man should venture to plant without giving due attention to the subject, and availing himself of all the experience of his neighbors; for experience, after all, is the only truly reliable guide.

2d. *The Soil.*—Having treated already of the different characters and modes of amelioration of soils, it is only necessary here to point out what particular qualities or kinds are best adapted to the different classes of fruit trees, as far as experience will warrant in so doing. There are soils of a certain texture and quality, in which, by proper management, all our hardy fruits may be grown to perfection. For instance, the soil of our specimen orchard, which is that usually termed a *sandy loam*, with a *sandy clay subsoil*, so dry that it can be worked immediately after a rain of twenty-four hours. On this we have apples, pears, plums, cherries, peaches, apricots, and, indeed, all the fruits planted promiscuously, side by side, not by choice but necessity, and all these yield bountiful crops of the finest fruit every season, and that, so far, without any special attention in the way of manures or composts. Our country abounds in such soils, and others somewhat different in character, but equally eligible for all fruit trees when well managed. On the other hand, there are soils wholly unfit for fruit trees of any kind—such are peaty or mucky, and damp, cold, and spongy soils. For an orchard of apples or pears, a dry, deep, substantial soil, between sandy and a clayey loam, and possessing among its inorganic parts a considerable portion of lime, is, according to all experience, the best.* On such soils we find the greatest and most enduring vigor and fertility, the healthiest and hardiest trees, and the fairest and best-flavored fruits. Trees both of apples and pears, planted on such soils in western New York, upwards of fifty

---

* The ashes of the bark of apple trees disclose the fact, that in one hundred parts upwards of fifty are lime. In the sapwood eighteen of lime, seventeen of phosphate of lime (similar to bone earth), and sixteen of potash. In the heart or perfect wood, thirty-seven of lime. In the ashes of the sapwood of the pear of one hundred parts, twelve of lime, twenty-seven phosphate of lime, and twenty-two of potash. In the ash of the bark, thirty of lime.

years ago, are, at this day, in the very height of their vigor and productiveness, without having received more than the most ordinary culture. In some of these soils, where the pear and apple flourish so well, and endure so long, the peach does not succeed at all. The reason is, it is too stiff and compact.

The plum succeeds best, as a general thing, on a clayey loam, rather stiff. The Canada or native plum, however, succeeds well on very light soils. The *cherry*, the *peach*, *apricot*, *nectarine*, and *almond*, require a light, dry, and warm soil, and will not succeed on any other. The best and most enduring peach orchards are on dry, sandy loams; but good orchards are raised with proper management on loose, light sands, though on such the trees are shorter lived, and require constant care in the way of dressings of manure and compost. There are two points to be observed in regard to soils under all circumstances. They must possess the inorganic substances, such as lime, potash, etc., that constitute a large portion of the ashes of the wood and bark when burned; and a sufficient amount of organic matter, vegetable mould, which dissolves and furnishes material for the formation and growth of new parts. When large and permanent plantations are to be made, it will well repay the trouble and expense of procuring the analysis of the soil, in order to ascertain somewhat correctly its merits and defects. People who have been long engaged in the culture of the soil, can judge pretty correctly of its quality by its appearance, texture, subsoil, and the character of the rocks and stones that underlie and prevail in it; but the inexperienced do not understand such indications, and will do well to have recourse to a careful analysis by some competent person.

3d. *Preparation of Soil for an Orchard.*—The season before planting, the soil should be at least *twice ploughed* with a common and subsoil plough, enriched with suitable composts, and drained, if necessary. It should be eighteen inches to two feet deep, and quite dry.

4th. *Enclosures.*—Before a tree is planted, it is necessary that the ground be enclosed with a fence, sufficient to protect it against the invasion of animals. It is no uncommon thing to hear people regret that the cattle broke into the orchard and destroyed many trees. Indeed it frequently happens that more damage is done in this way than, if duly estimated, would have fenced the whole orchard. There is much inquiry now-a-days on the subject of fences, and various plans and materials are suggested and tried. *Live hedges* are unquestionably the most ornamental and appropriate enclosures for extensive plantations of fruit trees, and in time will no doubt be generally adopted. Hitherto the failure of many plants tried, and the cost and difficulty of obtaining others, have retarded their introduction. Experience, however, has at length pretty fairly decided that the *Osage orange is the best for the west and south west,* and the *buckthorn for the north and east.* The seeds of both these plants are now easily procured, and plants of them may be obtained in nurseries at $5 or $6 per 1000, and about 2000 will fence an acre of ground, setting the plants twelve inches apart in two rows six inches apart, which is the strongest way. A single row at six inches apart will make a good fence with proper shearing to thicken them at the bottom; either way they will make a beautiful and efficient hedge in five or six years. The *honey locust* is also a strong, hardy, rapid growing plant, and makes a hedge in three or four years that animals will be *afraid to look at.* It is sometimes objected to hedges that they harbor birds, but it is to be remembered that birds are the natural foes of insects, and never fail to accomplish a vast amount of labor for the good of the fruit grower, for which they ought to be fully entitled to a participation in his enjoyments. As the feathered race are persecuted and driven away from our gardens, insects become more numerous and destructive; at least this is the experience of most people, and should lessen, if not entirely prevent, the cruel hostility that is continually waged against them.

5th.—*Selection of Varieties of Fruits for an Orchard.*—This is a most important point; the selection of varieties must in all cases be made with reference to the uses to which they are to be appropriated. The family orchard of the farmer, we will suppose to contain apple trees alone, as all the other fruits are, or ought to be, grown in the *fruit garden.* His selection of varieties must be adapted to his wants and circumstances. In the first place, the number of his family must regulate the proportion of kitchen and table varieties. In the second place, he must consider how many he will want for *sauce*, how many for *baking* and *drying*, how many for *cider*, and how many for the *dessert*, and what proportion of *sweet* and of *acid*. These are all considerations that depend upon the habits, taste, and mode of living of families, and for which no man can provide, or suggest, but the planter himself. Then, again, he must consider to what extent it may be advantageous to feed apples to his stock, and provide for it accordingly.

Without considering well all these points, a man may sit down and select what are called "the best varieties," and yet find himself badly suited when they come to bear; for so it happens that a variety that may be *best* for the dessert will be exceedingly unprofitable for other purposes. A hardy, vigorous, and productive variety of medium quality, quite unfit for the table, may be infinitely more advantageous for feeding stock, than a feeble growing, shy bearing variety, quite indispensable for the dessert; and an apple may be excellent for sauce, for baking, or drying, and unfit for the dessert; these points should all be duly considered.

The *Market* or *Commercial Orchardist* must exercise the same discrimination in the selection of his varieties, adapting them to the mode of culture he intends to pursue and the market he intends to supply. In the immediate vicinity of large cities and towns, where the orchardist may carry his fruit to market in a few hours, the most profitable culture will, generally speaking, be summer and early autumn fruits, or such as require to be consumed immediately after maturity, and are unfit for distant transportation. *Early* apples and pears only will be profitable for him, because the autumn and winter varieties can be sent so easily from the most distant portions of the interior with such facilities as our present system of railroads, plank roads, canals, and steamboats afford. In addition to early apples and pears, his position gives him great advantages for the profitable culture of all the *stone fruits, gooseberries, currants, raspberries, grapes,* and such soft fruits, when intended to be disposed of in a raw state.

The Market grower of the interior will find his most profitable culture to be principally, *autumn and winter apples and pears,* to which he may add *quinces;* because all these can be packed and transported to a great distance with safety, and the comparative cheapness of his lands enables him to compete advantageously with those more favorably situated in regard to market. He can only cultivate the summer fruits with a view to drying or preserving, or for the supply of a local demand. All orchard fruits, intended for profitable orchard culture, should be *first,* in regard to the trees, *hardy, vigorous,* and *productive.* The fruits should be of *good size, fair appearance, good keepers,* and of *good quality.* It should be borne in mind that many of the very best fruits are very unprofitable for *general* market culture. Under certain circumstances this may not be the case, as for example, in the neighborhood of such a city as *London*, or *Paris*, or even *New York* or *Boston*. A class of people are to be found in such places, who will pay almost any price for *extra fine* fruits. Where apples can be sold for $2 per bushel, pears at $1 per dozen, grapes at $1 per pound, and other fine fruits in proportion, growers are warranted in cultivating very choice sorts, even if they be difficult to manage and comparatively unproductive. As a general thing, however, taking the markets as they are, the great bulk of consumers preferring fruit of tolerable good quality and moderate prices, to the very best at twice or three times the ordinary price, the most profitable varieties will be those that can be produced at the least expense, provided always that they be *good;* for fruits of a decidedly inferior quality, whatever may be their other merits, are wholly unworthy of cultivation for the market. Another thing is the selection of varieties that succeed best in the locality where they are to be cultivated. A variety that succeeds remarkably well in any particular locality should, other things being nearly equal, be cultivated largely. The *Newtown pippin* apple, for instance, is a profitable orchard fruit on Long Island and on the Hudson, but in Western New York, no system of management would make it yield one-fourth as much net profit as the *Northern Spy, Rhode Island Greening,* or *Roxbury Russet.* Large plantations, for profit, should always be made up of well proved varieties, that have been tested in the locality, or one similar in regard to soil and situation. A list of select varieties will be given in a succeeding and separate part of the work.

6th. *Selection of Trees.*—For the farmer's orchard, where the ground among the trees is to be cultivated mainly with the plough, and occasionally cropped, standard trees, with stems four or five feet in height, will be the most eligible, and ought to be at time of planting three or four years old from the bud or graft, well grown, with stout, straight, well proportioned trunks. Low, stout trees are always preferable to tall, slender ones. Inexperienced planters are generally more particular about the height than the diameter of the trunk, but it should be just the other way. If trees are stout, and have good roots, a foot in height is comparatively unimportant, unless to one who wishes to turn cattle into his orchard and have the heads of his trees at once out of their way. Few people, however, follow such a practice. In very elevated and exposed situations low trees are to be preferred, as the wind does not strike them with such force as it does the tall ones.

7th. *Arrangement of the Trees.*—The distance between the trees in an apple orchard should be thirty feet from tree to tree in all directions. In a very strong and deep soil, where the trees attain the largest size, forty feet is not too much, especially after the first fifteen or twenty years. There is a great difference between the size that different varieties attain, and in their habits of growth. One will attain nearly double the size of another within ten years. Some are erect in their habits (as fig. 3); others spreading (as fig. 5); and it will add greatly to the symmetry of the plantation, if the trees of the same size and habit of growth be planted together. Varieties that ripen about the same time should also be planted together, as the maturity can be more easily watched and the fruit gathered with much less inconvenience. The largest fruits being most liable to be blown off, should be placed in the least exposed quarter.

The ordinary arrangement of orchard trees, is the square or regular form, in rows the same distance apart, and an equal distance between each tree. Thus, in planting a square of one hundred feet, for example, the trees to be twenty-five feet apart, we commence on one side, laying a line the whole length. On this line we measure off the distances for the trees, and place a stake indicating the point for the tree. Thus, in fig. 93, we have five rows of five trees each, making twenty-five in all, and all twenty-five feet apart. This is the simplest, and probably the best for very small orchards. The better plan for large orchards is what is called *quincunx* (fig. 94), in which the trees of one row are opposite the spaces

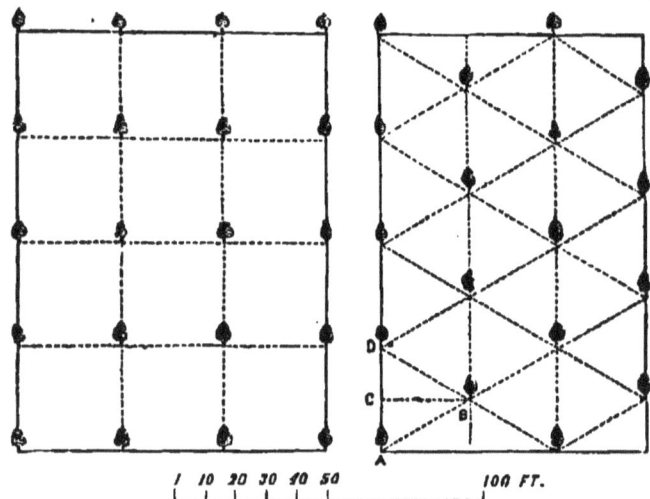

Fig. 93, square planting.   Fig. 94, quincunx planting.

in the next. In this way, although the trees are at equal distances, there is a larger clear area around each tree. In fig. 94, the square form, every tree stands in the corner of a square in the centre of, and equally distant from *four* others. In the quincunx, every tree stands in the angle of a triangle of equal sides, and in the centre of, and equally distant from *six* others. Thus, in the latter, there is a greater space left for the admission of light and air, and trees so planted may be at less distances than in the other. The operation of planting is more complicated than that of the square, the rows not being the same distance apart as the trees are in the row. The first thing to be done is to find the two measures. Suppose, for instance, we propose to plant a plot of ground one hundred feet square, and to have the trees twenty-five feet apart every way, we make a triangle of wood, A, B, D, each side of which is twenty-five feet; we then measure the distance from the angle B to the centre of the opposite side at C, and this gives us the distance between the rows, which will be about twenty-one feet. This will be called the *small* measure; and with this we measure off on two sides the distances for the rows, and put down a stake at each. We then commence on the first row, and with the long (twenty-five feet) measure mark off the places for the trees, and put down a stake to each. The measurements must be made with exactness, in order to have the plantation present a regular appearance, as in fig. 94.

8th. *Selection of Trees for the Market or Commercial Orchard.*—The remarks made in reference to the selection of standard trees for the family orchard, may be applied with equal propriety to these; but the orchardist must be supposed to have invested a considerable amount of capital, and probably devotes his entire attention to his trees, and depends upon them for his support. It is, therefore, a great object with him to have early returns in the form of products. An orchard of standard apples will not produce any considerable quantity of fruit before the eighth or tenth year, nor pears before the twelfth or fifteenth year. In the mean time, it is highly desirable to occupy the ground amongst the trees in some way that will at least bear the expenses of cultivation. If this can be done, it is as much as can be expected in the usual practice of cultivating root crops. The most profitable manner of turning to account the spaces between the standard trees for the first ten or twelve years at least, is to plant them with dwarf and pyramidal trees, or dwarf standards, that will commence bearing the third or fourth year after planting. This is the course pursued by the orchardists of France and Belgium, where land is valuable, and the cultivators are compelled to turn every inch of it to the best account. Attention has been slightly called to this mode of management in this country, and a few persons have already carried it into practice. As soon as it comes to be considered, it cannot fail to recommend itself to those who are embarking extensively in the orchard culture of fruits for the market, on high-priced lands. It is only surprising that it should have been so long overlooked by shrewd and enterprising orchardists. An acre of land, for example, planted with standard apple trees, at thirty feet apart, contains forty-five to fifty; and if we fill up the spaces with *dwarfs* on paradise, at six feet apart, leaving ten feet clear around each standard, we get in about five hundred dwarf trees. These will bear the third year, and during the next five years the average value of their products will be at least twenty to fifty cents each. We would plant them in such a way that the plough and cultivator could be used among them, two dwarfs between each standard, and two full rows between each row of standards, as in fig. 95.

In very rich and deep soil, when it may be necessary to give the standards thirty-five or forty feet, there may be two pyramidal, or low standards, on the Doucain stock between two standards, and one row of pyramids and two rows of dwarfs between two rows of standards

THE ORCHARD. 171

In seven or eight years the dwarfs might be taken out, and the pyramids remain till the twelfth year.

Orchards of standard pears may, in the same manner, be filled up with dwarf and pyramidal trees on the quince. Standard pears do not require so much space as apples, their branches generally are more erect. In this country

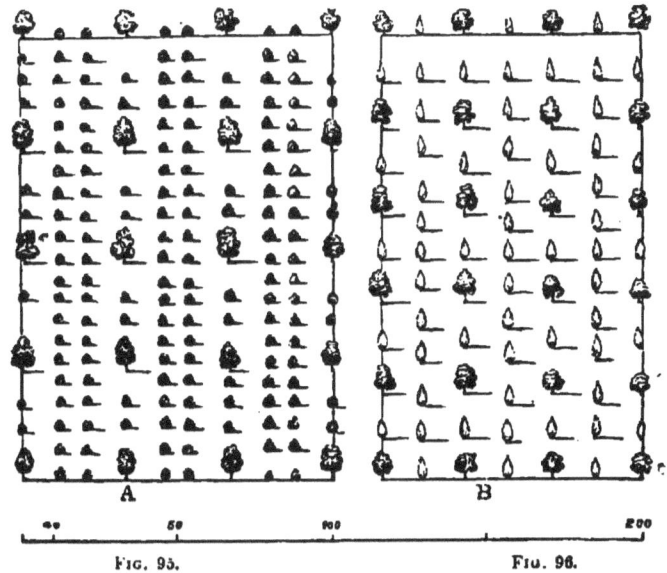

Fig. 95. Fig. 96.
Fig. 95. orchard of standard and dwarf apple trees. *Fig* 96, orchard o standard and dwarf or pyramidal pears.

standard pears should not have naked trunks over four feet high at most, and twenty-five feet apart is quite sufficient; at this distance an acre will contain about seventy trees. These, as a general thing, will not begin to bear until the tenth year, unless artificial means be resorted to. By putting one pyramid, or low standard, between each in the same row, and a row ten feet apart between each row of standards, as in fig. 96, we can plant 250 dwarfs, or pyramids, that will commence bearing the third year, and will be in full bearing the fifth; yielding not less on an average than $1 to $2 per tree.

To give trees a perfectly pyramidal form requires considerable care and skill in their management. This will be spoken of presently in treating of the fruit garden; but very beautiful and prolific low standards may be made on the quince, with stems about two feet high, and the heads above that point left to branch in their natural way. Trees of this form bear full as soon as the pyramids, because they are pruned less; they may always be relied upon for a crop the second or third year after planting. We have gathered upwards of fifty large and perfect specimens from trees four years old, and many had been thinned off. Trees of the white Doyenne have produced upwards of twenty very large specimens the third year, from the bud. Fig. 97 is a portrait of a four-year-old Louise Bonne de Jersey, on quince, never pruned.

Fig. 97.
Half standard pear tree on quince.

In selecting pears on the quince for profitable orchard culture among standards, varieties should be chosen that succeed particularly well on the quince, such as *Louise bonne de Jersey, Duchess d'Angouleme, Beurre, Diel, Bartlett, White Doyenne, Vicar of Winkfield, Glout Morceau, Easter Beurre*, &c., &c. All these, and many others that will be named hereafter, grow vigorously, bear early, and produce larger, and in all respects finer fruit on the quince than on the pear. S. B. Parsons, Esq., of Flushing, Long Island, of the well known nursery firm of Parsons & Co., has planted an orchard of four acres with 440 standard pears at twenty feet apart, and among these he planted pears on quince ten feet apart, which gives him 1320, making the whole number 1760 on the four acres. The ground he selected was an old pasture with a light loamy soil, but not inclining to sand, and a subsoil of hard pan. This he planted with corn until the ground was well mellowed, and then put in two sloop loads, or 3000 bushels of stable manure, worth on the ground $175. The first year after planting he cropped the orchard with corn, but found it injurious to the trees; since that he has cropped it with potatoes and sugar beets alternately, and with good management these can be made to pay for the manure, and sometimes the labor. All those on their own root, except one row, are the *Lawrence*, a native Long Island variety, and those on the quince the *Glout Morceau, Vicar of Winkfield, Louise Bonne de Jersey, Winter Nelis, Lawrence,* and *Beurre d'Arremberg*. He adds, that at the time of writing, December 10, 1850, some of the Vicar of Winkfield trees planted in 1849, had fifty to seventy-five fruit buds each, and expects them to produce the fifth year from planting, *one dollar* per tree. Within the past two years, several extensive plantations, wholly of pears on the quince, have been made, and considering the quick return they yield, their prolific nature, and the number of trees that can be planted on a small space of ground, they cannot fail, under good management, to prove highly profitable. *Peach trees* should be thrifty yearlings that have not been pruned up during the summer, the side branches having been shortened only, and regulated by pinching. At the time of planting they may be pruned up, so as to leave three feet of a clean stem. They may be set at the distance of fifteen feet, and even twelve will be found quite sufficient, if the heads are annually pruned, as will be directed hereafter. The peach grows so rapidly, and commences to bear so soon, it would not be advisable to plant any fruit trees in the spaces, unless currants or gooseberries; a row or two of which might be put between two rows of the peaches for the first four or five years after planting. *Standard cherries* on mazzard stocks should not be over two years old from the bud, with stems five feet high. In the west and south, where the trees are subject to the bursting of the bark on the trunk, it is advisable to have the trees branched as near the ground as possible; and in such cases the *Mahaleb stock* is better than the mazzard, as it makes lower, more compact, and fertile trees. Orchards of pyramidal, or low dwarfs, on the Mahaleb may be planted at twelve feet apart, or the ground may be more compactly filled by planting standards and dwarfs alternately, as in the case of the pears.

Apricots on peach stocks may be planted in the same soil, and should be of the same age and character as the peaches. On plum stocks they are better adapted to heavy soils. Plum trees for orchard standards should be about two years old from the bud or graft, with stems about three feet high. The stone fruits in particular should have low stems, as they are more subject to the gum on the trunk if pruned up high. They may be planted at fifteen feet apart, the same as peaches and apricots. Quinces should be two years old at least, and may be three from the layer, cutting, or bud, with a stem two feet high, clear of branches: they may be planted twelve feet apart, which gives about 300 to the acre.

9th. *Pruning and Preparing the Trees for Planting.*— When a tree is taken up from the nursery, it unavoidably loses some of its roots, and others are more or less mutilated; the roots frequently suffer, too, by long carriage or exposure, and in this state it is unable to support the entire head as it came from the nursery. This has been previously explained. In order that a tree may grow, it is necessary that a balance should exist between the stem or branches and the root; consequently, when a tree is transplanted, its branches should be reduced by shortening so as to correspond with the roots. A standard tree that has four or five branches forming a head, should be pruned at the time of planting to within three or four buds of the base of each of the branches. These remaining buds, receiving all the nourishment, will push vigorously; whilst if the branches had been allowed to remain entire, they would have required a greater supply of food than the roots could have furnished, and the tree would either have died or made a very feeble growth. Every bud we leave on the top of a tree, will produce either leaves or shoots, and these are so many new individuals requiring sustenance. If we leave on one hundred, it is plain the demand will be much greater than if we leave only twenty. The roots must be dressed by cutting back all bruised points to the sound wood, with a smooth cut on the under side of the root. Trees thus prepared are ready for planting.

10th. *Planting Orchards.*—When the soil has been thoroughly prepared by subsoil ploughing, or trenching and manuring the season previous, the planting is a simple matter, but if this has not been done, planting properly requires considerable labor; for large holes three or four feet wide and two feet deep must be dug for the trees, and the requisite composts procured to be mixed with the earth in which the roots are to be placed. Whatever manures be applied at this time should be perfectly decomposed; as, if fresh and warm, they will burn the roots. Trees are often killed in this way. The planting offers an excellent opportunity for supplying any defects in the soil; for instance, if too compact, sand, leaf mould, muck, &c., may be added to render it more porous; and if too light, clay, stiff loam, ashes, &c., may be added to make it more retentive. The proper way to furnish these materials is to dig large holes and put a good bed, twelve to eighteen inches deep, of the compost in the bottom under the trees. Lime should form a part of all composts, and especially for the apple and pear; half a peck may be mixed with the bed of each tree in soil not naturally calcareous. In digging the holes, the good surface soil should be laid on one side, so that it can be used to fill in among the roots, and for this purpose it should be as finely pulverized as possible.

When the compost has been laid in the bottom of the hole, and a layer of fine surface soil spread over it, so as to be highest in the centre, the tree is set on it, so that when the planting is finished, the collar will be about two inches below the surface. In the case of trees on dwarf stocks, such as pears on quince, *all the stock* must be under the ground. The roots must be carefully adjusted so that each one is spread out in its natural position; the fine earth is then filled in amongst them so that no vacancies will be left; the upper roots should be held back by the person who holds the tree until the lower ones are covered. When the filling in is half done, it may be gently trodden down with the foot, so as to give the tree a firmer hold of the ground. In advanced spring planting, a pail of water might be given to each tree when the earth is filled partly in; at other times it is unnecessary, if not injurious.

11th. *Staking.*—Where the trees are large, or the situation is exposed, either one or two stakes should be planted with each tree, to which it must be kept fastened for the first season, until the roots have fixed themselves in the ground. A proper provision must be made to prevent the tree from rubbing or chafing against the stake. When two stakes are used it may be fastened to each in such a way as not to rub against either.

12th. *Mulching.* This should be looked upon as an

indispensable operation in all cases. It consists in laying on the surface of the ground, around the trees, to the distance of three feet or so, a covering of half decomposed manure, saw dust, spent tan-bark, &c., two or three inches deep. This prevents the moisture of the soil from evaporating, and maintains a uniformity of heat and moisture which is highly favorable to the formation of new roots. It also prevents the growth of weeds around the tree, and obviates the necessity of hoeing, dressing, or watering, during the season. We frequently practise it among nursery rows of late spring-planted trees with great advantage. A deep mulching should always be given to fall-planted trees to prevent the frost from penetrating to the roots or drawing up the tree.

13th. *After-management of Orchard Trees.*—This consists in the cultivation of the soil among the trees, and pruning them to regulate their growth. For the first five or six years after planting, the ground among orchard trees may be advantageously cropped with potatoes, rutabagas, or sugar beets. The manuring and culture that these roots require, keep the soil in good condition, and will assist in defraying the expenses of the orchard. Grain crops should never be planted among trees, as they deprive them of air to a very injurious extent. If no root crops are cultivated, the ground should be kept clean and mellow with the one horse plough and cultivator, the same as recommended for nursery culture. Every third or fourth year, the trees should receive a dressing of well-decomposed manure or compost adapted to the wants of the soil and the tree, worked in around the roots with the forked spade. This should always be done in the fall. Dwarf apples and pears require more frequent and liberal manuring than standards, because their roots occupy a limited space; their heads are large compared with the roots, and they bear exhausting crops. Whoever has a large plantation of these trees, should be well provided with heaps of compost a year old, and give each tree a peck to half a bushel before the setting in of winter every year. This will maintain their vigor, and ensure large and regular crops of fine fruit. Directions for pruning and forming the heads of standard trees, will be treated of under the general head of pruning.

Section 3.—The Fruit Garden.

The fruit garden is a plantation of fruit trees intended to supply the family with fruit. In some cases, where a large supply of fruit is wanted, and the proprietor has land and means to warrant it, a certain portion of ground is wholly devoted to it; and in others, it forms a separate compartment of the kitchen garden, or is mixed with it— the fruit trees occupying the borders or outsides of the compartments, and the culinary vegetables the interior. The latter is most general, in this country, at the present time. In a country like ours, so well adapted to fruit culture, where almost every citizen of every rank and calling not only occupies but owns a garden, and, as a general thing, possesses sufficient means to enable him to devote it to the culture of the higher and better class of garden productions, the fruit garden is destined to be, if it is not already, an object of great importance. In the old countries of Europe, the rich alone, or those comparatively so, are permitted to enjoy such luxury; for land is so dear that working people are unable to purchase it, and if they are, they are either unable to stock it with trees, or their necessities compel them to devote it to the production of the coarsest articles of vegetable food that can be produced in the greatest bulk. It is not so in America. Here every industrious man, at the age of five-and-twenty, whatever may be his pursuits, may, if he choose, be the proprietor of a garden of some extent, and possess sufficient means to stock it with the finest fruits of the land.

The present actual state of the population gives abundant evidence of this happy and prosperous condition. Let us look at our cities and villages. In Rochester, excepting a narrow circle in its very centre, every house has its garden, varying in extent from twenty-five by one hundred feet to an acre of ground; and not one of these but is nearly filled with fruit trees; and so it is, but on a larger scale, in all the villages of western New York—a section of country in which the first white man's settlement can scarcely date back over fifty years. Aside from the beneficial results to individual and public health and prosperity from this general union of the fruit garden and the dwelling, it cannot fail to exercise a softening and refining influence on the tastes, habits, and manners of the people, and greatly strengthen their love of home and country.

The great thing wanting at this moment, is a knowledge of the correct method of planting and managing fruit gardens. We cannot pass along the streets a rod, where there is a garden, without seeing and feeling that three fourths of the profit and pleasure which gardens might afford, are sacrificed to bad management, arising, in the main, from ignorance of the proper modes of culture adapted to such limited grounds; and it is hoped that the suggestions and plans offered in the following detail of fruit garden management, may afford at least a portion of the information wanted.

The formation of a fruit garden requires a consideration of the *soil, situation, enclosures, laying out, selection of trees, selection of varieties,* and *planting.*

1st. *The Situation.*—This is generally governed by the particular circumstances of the proprietor, those only who build with reference to the location of the garden, or who have a large domain at their disposal, having an opportunity of selection to any considerable extent. Persons who live in cities and villages, have to make the best of their situation. As it is, if it be exposed, they can only give it protection by lofty enclosures, that will break the force of the winds. The *aspect* they cannot alter, and must adapt other circumstances to it. Those who can should select a situation convenient enough to the dwelling, to render it at all times easy of access, in order to save time and labor in going to and from it. It should also be sheltered from the north and west winds. The former are destructive to the blossoms in spring, and the latter frequently blow off the fruit before its maturity. In sections of the country subject to late spring frosts, an elevated situation is to be preferred, as in the case of orchards. A full eastern or southern aspect should be avoided, because in them the sun's rays strike the trees while the frost is upon them, and produce injuries that would be avoided in other aspects. Where artificial shelter is required, a belt of rapid-growing trees, composed of evergreens and deciduous trees mixed, should be planted on the exposed side, but at such a distance as to obviate any difficulty that might arise from the injurious effects of shade, or from the roots entering the garden. Such a belt of trees might, at the same time, be made to impart a pleasing and highly ornamental appearance to the grounds.

2d. *The Soil* is a most important consideration. As in a garden a general collection of all the fruits is to be grown, and that in the highest state of perfection, the soil should be of that character in its texture, depth, and quality, best adapted to general purposes. It should not only be suitable for the apple and the pear, but for the peach, the cherry, and the plum—a good, deep, friable loam, with a gravelly clay subsoil, and entirely free from stagnant moisture. In this country, our warm summers, and frequent, protracted droughts, render a deep soil for a garden absolutely necessary. *Two feet* is little enough, and three would be still better. The means for deepening, drying, improving, and changing the character of soils have been already pointed out under the general head of soils, and need not be repeated here. Suffice it to say, that it will always be found true economy to be liberal in the first preparation of the soil; for after a garden is laid out and permanently planted, improvements are always made with greater difficulty and expense.

Enclosures.—The cheapest and most ordinary kind of enclosure for gardens in this country, is the tight board fence, and the picket or paling fence. The former should be made of stout cedar posts, set at six feet apart, and three or four feet in the ground, the ends being previously charred to increase their durability, connected in the middle and on the top with cross-bars or rails which may be two by four inches. The boards should be well seasoned, matched, and securely nailed to the cross-bars. Where the fence is required to be higher than the posts, the boards can extend above the top rail two, three, or even four feet, if necessary. The picket or paling fence is made in the same way, as far as the framework, posts, and cross-bars go; but, instead of matched boards, pickets, from three to six inches wide, and pointed on the top, are used, and a space of two inches left between each. Where the proprietor can afford the expense of a brick or stone wall, it will prove the most permanent, and, in the end, the cheapest enclosure. The height of the fence or wall depends somewhat on the extent of the garden. In ordinary cases, eight or ten feet is the proper height, but when the garden is very small, five or six feet is enough; and the open paling will be preferable except on the north side, to the tight board fence, as it offers less obstruction to the air and light. A high fence around a very small garden, besides being injurious to vegetation in it, looks quite out of character, giving to it the appearance of a huge box. Live hedges, as recommended for orchards, might be employed around country gardens of considerable extent, say an acre or upwards, but they require to be kept in the neatest possible condition.

Trellises.—In England, and other parts of Europe, where the summer temperature is not so high as it is here, espalier trees are trained directly on the garden walls or fence ; but our hot sun renders this unsafe, except in the case of the grape, or on the north sides of the walls. The sun strikes the south side of a fence with such force that the foliage in contact with it is burned. It is therefore necessary, where the walls or fences are to be occupied with espaliers, to erect suitable trellises at the distance of six to twelve inches from them, on which to train the trees ; the form of these differs according to the nature of the subject to be trained. They are generally made of upright and cross bars, of inch boards three inches wide, placed within six to twelve inches of each other, according to the growth of the species ; the larger the foliage and the longer the shoots, the greater may be the distances ; thus, the grape twelve inches, and the peach eight. Sometimes they are constructed of wooden bars and wire rods alternately ; these answer a good purpose for the grape, as it fixes itself to the wires by the tendrils. The trellis is fastened to the wall by iron hooks, and should stand a little farther from it at the bottom than at the top, for the purpose of giving the tree a better exposure to the sun, rain, &c. Fruits are grown so successfully in this country in the open ground that walls or trellises are seldom used, except to economize space. In the north, however, where the more tender fruits do not succeed in the open ground, walls may be advantageously employed, as the trees trained on them are easily protected both from winter and spring frosts.

Laying out the Fruit Garden.—This is the arrangement or distribution of the ground into suitable plots or compartments, necessary walks, etc. The mode of doing this depends on the size of the garden, and the manner in which it is to be planted. Fruit gardens, properly speaking, are such as are wholly devoted to fruits; but a very common form, as has been already observed, is the *mixed* garden, where a portion only is devoted to fruits, and the remainder to culinary vegetables. We will first consider

The Fruit Garden proper.—In all fruit gardens the number of walks should be no greater than is absolutely necessary for convenience. In small places the better plan appears to be, to carry the principal walk around the outside, leaving as much as possible of the interior, where air and light are enjoyed to the greatest extent, for the trees. A border should be left between the fence and the walk, of sufficient width for the trees to be trained on the fence trellis. If appearances were to be strictly observed, this border should be as wide as the fence is high, but as a general thing five to six feet will be sufficient; and where ground is limited, appearance must in many cases be sacrificed to economy. Where the work is all performed by manual labor, the walks need not be more than five to six feet wide, as that admits of the passage of a wheelbarrow; and this is all that is required.

Fig. 98 is a design for a very small garden fifty feet by one hundred. *A* is the entrance gate, four feet wide: *B*, *B*, a walk five feet wide; *C*, *C*, fence border, six feet wide. The rows of trees are eight feet apart. The pyramidal pears and cherries, Nos. 1, 2, 3, and 4, at seven feet apart in the row. Nos. 5 and 6, dwarf apples, at four feet apart. No. 7, pyramidal or dwarf standard plums, at seven feet. Nos. 8, 9, and 10, low standard peaches, at ten feet apart, the outside ones four feet from the walk. Nos. 11, 12, 13, and 14, low standard quinces, etc. Nos. 15, 16, 17, 18, 19, and 20, espaliers, apricots, grapes, etc.

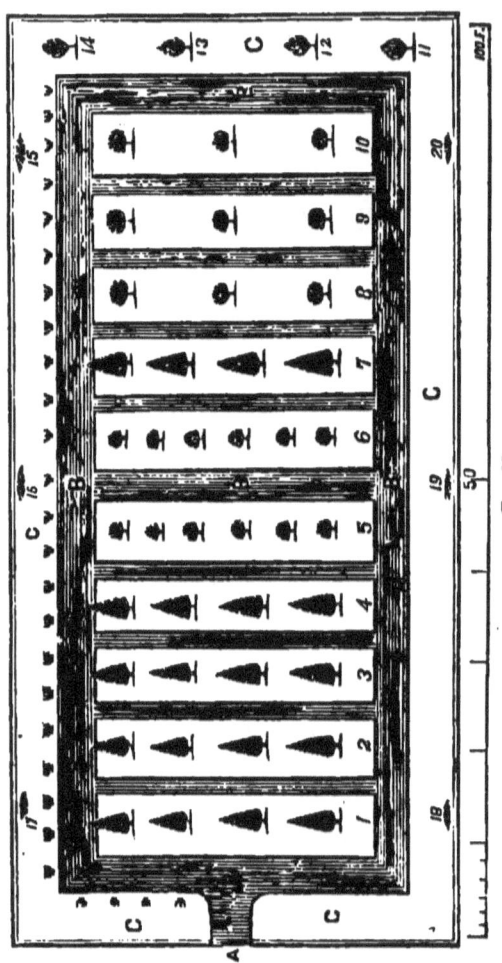

Fig. 89.

One border is filled with gooseberries and currants, the other can be occupied with raspberries and strawberries. This arrangement gives in this little garden twenty pyramidal trees, thirteen standards, twelve dwarfs, six espaliers, besides space enough for two dozen currants, two dozen gooseberries, two dozen raspberries, etc. For several years a few strawberries and low vegetables, such as lettuce, radishes, beets, carrots, turnips, or even *dwarf* peas, may be grown in the spaces among the trees, but in no case to be permitted nearer than within three feet of the tree.

A walk through the centre would be necessary, and this should be ten feet wide, and there should be a turning place left at the end opposite the entrance.

The mixed, or fruit and kitchen garden, is laid out in a similar manner; the trees are planted in rows on a border six to ten feet wide, according to the size of the trees, along the walks, leaving the interior of the compartments for vegetables. This arrangement is a very common one, and generally answers a very good purpose; but where it is practicable, it is much better to devote a separate portion exclusively to fruit, in order that the one may not in any way interfere with the other. In such a garden, the number of the walks, and consequently fruit borders, will depend upon the proportion of the ground intended to be allotted to fruit, and this again will be regulated by the means, tastes, and demands of the family.

Fig. 99 (see frontispiece) is the plan of a mixed fruit and kitchen garden, one hundred and fifty feet wide by two hundred long, being one hundred and ten square rods, somewhat less than three quarters of an acre. The design is to have two tree borders exclusive of the outside or fence border. The centre main walk from A to C, is ten feet wide. That crossing it in the centre six feet wide. The small walk next the fence border four feet wide, and that between the two tree borders five feet. The fence border is six feet wide, and may be planted with espalier trees, vines, etc., besides currants, raspberries, strawberries, or anything of low growth, not requiring the fullest exposure. The tree borders are all eight feet wide, except the dwarf apple border, which is only six. The outside border is planted on the two sides with low standard peaches, apricots, plums, quinces, etc., at twelve feet apart, and the two ends with pyramids at eight feet.

The inside borders are planted with pyramids and dwarfs, the former at eight, and the latter at six feet apart. *A*, is the entrance; *B*, well or cistern; *C*, a space to turn a horse and cart upon. This arrangement gives thirty standard trees, eighty-three pyramids, and forty dwarfs, leaving clear the outside border over six hundred and sixty feet long and six wide, and the four interior compartments each about thirty by sixty feet. In cropping the latter with vegetables, they may be divided as in the design into narrow beds three or four feet wide, separated by paths eighteen inches wide.

Walks in the Fruit Garden.—The number of these, as has been remarked, should be simply sufficient for conducting the operations of gardening with convenience; this being provided for, the fewer the better. Where horse labor is employed, the main walk, either through the centre or around the sides, should be nine or ten feet wide. Where manual labor alone is employed, as in small gardens, five or six feet will be sufficient, and even four feet, as that admits of the passage of a wheel-barrow. Between each compartment, or line of trees, there should also be a path two or three feet wide, as a passage for the gardener or workmen, and others who may desire to inspect the trees. Where the expense can be afforded, the mains walk should be gravelled so as to be dry and comfortable at all seasons and in every state of the weather; for it is presumed that every man who has a fruit garden, worthy of the name, will wish to visit it almost daily, and so will the members of his family and his friends who visit him. The labor and expense of making a walk depends upon the nature of the soil. If dry, with a porous subsoil, absorbing water rapidly, six inches of good pit gravel, slightly rounded on the top, will be sufficient. If the soil be damp, and the subsoil compact, it will be necessary to remove the earth to the depth of a foot in the centre, and rising towards the sides, so that the excavation will resemble a semicircle; this is filled with small stones and a few inches of good pit gravel on the top. This makes a walk dry at all times. We often see very comfortable and neat looking walks, made of spent bark from the tannery; six inches deep of this will last two or three years, and no excavation is necessary in any kind of soil. It is not to be supposed that so great expense will be incurred, in any case, in the formation of the walks of a fruit or kitchen garden, as those of a pleasure ground or flower garden, and, therefore, it is unnecessary to suggest either costly modes or materials. The chief point is to secure dry, comfortable walking, without introducing any material that will produce a decidedly unpleasant contrast with vegetation. This can all be accomplished by the cheap and simple means referred to, and others that may suggest themselves.

The main walks alone should be gravelled; the smaller alleys or paths between the different lines of trees or compartments of the garden are principally for the use of the workmen. In very small gardens, where it is important to economize the ground, the spaces devoted to the walks may be of plank raised up on pillars or blocks a foot from the ground; the roots of trees can then penetrate the ground below the walk as well as the border, and scarce any ground will be lost.

Water.—A supply of water in the garden is a most important consideration in our warm, dry, sunny climate. Good crops of culinary vegetables cannot be secured in many seasons without a liberal application of water, and fruit trees are greatly benefited by frequent showering, especially in dry weather. It refreshes them and drives away insects. A good well or cistern should therefore be provided in every garden, and be situated as near the centre as possible, to be convenient to all parts.

SELECTION OF TREES.

1st. *Their Form.*—We start upon the principle that, in all cases, tall standard trees, such as are usually planted in orchards, are totally unfit for the garden. This is the one great and universal defect in American fruit gardening. The trees for a fruit garden should be all either *dwarf standards*, with trunks two to three feet high, *pyramids*, branched from the ground, or *bushes* with stems six to twelve inches high. Trees in these forms are, in the first place, in keeping with the limited extent of the garden, and convey at first sight the idea of *fitness*. In the second place, they give a great variety on a small space, for three or four such trees will not occupy more space than one standard. In the third place, they are in a convenient form for management, they are easily pruned or protected, and the fruit is easily gathered and less likely to be blown off than on tall trees. In the fourth place, they bear several years sooner than standards, especially pears and apples.

Among the forms mentioned, the *pyramid* is certainly the most beautiful; and in the best fruit gardening regions of Europe, where almost every conceivable form of tree has been tried, it is to-day the most popular, because it has proved the most advantageous and successful. The *apple for pyramids* should be on the *Doucain* stock. Certain varieties, such as the *Hawthorndean, Keswick Codlin, Summer Rose, Duchess of Oldenburg*, and many other moderate growers and early bearers, will make good pyramids on free stocks, but they will require more summer pruning and careful management to keep their vigor under check than they would on the Doucain.

But apples for the fruit garden, even on the Doucain, should be such as naturally make small trees and are inclined to early bearing. In these respects it is very well known there is a wide difference between varieties. Those mentioned above, and others similar in character, frequently bear, on free stocks in the nursery rows, at the age of three or four years from the bud, whilst others do not bear until eight or ten years old. This is a point that should always be looked into in selecting garden trees, for it is the natural and proper desire of every one who plants a tree in the garden to obtain fruit from it as early as possible.

The Apple for Dwarfs.—The apple, worked on the paradise, makes a beautiful little dwarf bush. We know of nothing more interesting in the fruit garden than a row, or a little square, of these miniature apple trees (fig. 100), either in blossom or in fruit. Those who have not seen them, may imagine an apple tree, four feet high, and the same in width, of branches covered with blossoms in the spring, or loaded with magnificent golden and crimson fruit in the autumn. They begin to bear the third year from the bud, and the same variety is always larger and finer

Fig. 109.
Dwarf apple tree.

on them than on standards. We had *Red Astracans* on paradise the past season, that measured eleven inches in circumference. The French plant a square or compartment of these in the kitchen or fruit garden, as they do

gooseberries and currants, six feet apart, and call it the
' *Normandie;* they also alternate them with pyramidal
pear trees in rows; and in some of the best mixed kitchen
and fruit gardens, two dwarf apples are planted between
two pyramidal pears, thus giving double the number of
them as of the pears in a border or row. In small gardens the apple should not be admitted under any other
form, and even to a limited extent in that, for it is the
great fruit of the *orchard*, and in nearly all parts of this
country they are extensively grown, and can be purchased
at very moderate rates.

The Pear, as a Pyramid (fig. 101).—The pear is eminently *the* tree for the pyramidal form, either on the free stock, or on the quince; on the latter, however, the trees bear much earlier, are more prolific, more manageable, and consequently preferable for small gardens. On the pear stock they require constant summer pruning and pinching, and in some cases, root pruning, to subdue the natural vigor, and induce early fruitfulness. Certain varieties, however, do not succeed on the quince,

Fig. 101.
Pyramidal pear tree, 7 feet high—4 feet wide at the base.

but the majority of melting varieties do, and produce larger and finer fruit on it than on the free stock. The tardiness of bearing of the pear tree, when grown in the ordinary standard form on pear stock, has, more than any other cause, retarded its general cultivation. No better proof of this can be adduced than the general partiality now shown for trees on quince stocks, that bear at the age of three or four years. The introduction of these trees, a few years ago, was really the first thing that gave a general impulse to pear tree planting. With most people, it is a very important thing to obtain fruit in two or three years, instead of waiting eight or ten. The best management of trees on free stocks, cannot bring them into a bearing state short of six or seven years, unless it be some remarkably precocious variety. People, therefore, who wish pear trees for pyramids that are easily managed, and will bear early, will select them on quince stocks, in case the varieties they wish to cultivate have been proved to succeed well on it.

The Pear in the dwarf standard form, as in the pyramidal, is much easier managed, and bears much earlier on the quince than on the pear; indeed, these trees arc as easily managed as a standard apple tree. There are some dwarf standards on the quince in our grounds here, and in gardens in this city, that are now eight years old, and about seven to eight feet high, with trunks from two to three feet, heads four to five feet high, and three or four feet in width, that have borne regular and heavy crops for the last four or five years, without any other care than thinning out superfluous wood. *The Cherry* is as easily managed in the pyramidal form as the pear, not only the free-growing sorts, *Hearts* and *Bigurreaus*, but the *Dukes* and *Morellos;* the latter, however, are less vigorous, and more easily managed. All should be worked on the *mahaleb* stock; this has the same effect on the cherry, to a certain extent, as the quince has on the pear. After the second or third year's growth, it subdues their vigor, and induces fruitfulness. We have a collection of upwards of thirty varieties, of four to five years old, that are now fine pyramids, from five to eight feet high, and they have all borne since the third year, and we find them quite as easily managed as the pear. The Dukes and Morellos should be chosen, where very small trees are desirable, as they can be grown in bushes like the apple on the paradise stock, at five feet apart.

Fig. 102 is the portrait of a dwarf Florence cherry tree, given by Mr. Rivers, in his Miniature Fruit Garden, only two years old, bearing fruit. Our dwarfs frequently bear the third year.

Fig. 102.
Dwarf cherry, two years from bud, bearing.

The Plum as a Pyramid.—The plum has rarely been cultivated as a pyramid, but recent experiments prove that it is quite susceptible of that form under proper management. It should be worked on a stock calculated to subdue its natural vigor. The native or Canada plum answers a good purpose, the mirobalalan or cherry plum, and the sloe (prunus spinosa) dwarf it, to a still greater extent. Summer pruning and pinching, as well as occasional root pruning, are all necessary to check the vigor of most kinds, and keep them in suitable dimensions for small gardens where it is necessary to plant them close.

The Plum as a Dwarf Standard.—Besides the pyramid, this is the only form in which the plum should be admitted in the garden. The dwarf standard, with a trunk two or three feet in height, and a symmetrical round head, is a very pretty and appropriate form, and requires less skill and care in the management than the pyramid, and by proper management the trees require but little if any more space.

The Peach.—The best garden form for the peach is that of the *dwarf standard*, with a trunk eighteen inches to two feet. With proper management, which will hereafter be described, this form is easily conducted, even when the trees are on peach stocks. The plum stock, and especially the sorts recommended for dwarf plums, gives trees that are less vigorous and more easily kept in a small space. In nearly all parts of our country the fruit ripens perfectly in the open ground, so that espalier training, as has been remarked, is seldom practised, unless to save ground; or in northern localities, where protection of the buds during winter, or of the blossoms in the spring, is necessary. In such cases alone are espaliers to be recommended, as they require much greater care in pruning and training than in any other form. Espalier trees are of various forms, but the *fan*, as it is termed, is the best adapted to the peach. It consists of two main branches or divisions of the stem, spread out in the form of a V; each of these bears a certain number, as many as may be necessary to fill the trellis, of secondary branches, and these furnish the bearing wood. The production and management of this and other espalier forms, will be treated fully under the head of pruning and training.

The Apricot and Nectarine.—The remarks applied to the peach apply with equal force to both these trees; they succeed equally well as low standards, or as espaliers. The apricot is more generally grown in this form than any other tree, because its early blossoms are so easily protected, and the curculio does not appear to be so troublesome to it as in the standard form.

The Quince, in the garden, should either be a dwarf bush, with a stem twelve to eighteen inches high, and a compact, symmetrical head, or a pyramid. In the latter form it is quite easily conducted, but requires more care, of course, than as a bush, as the upper part of the tree must be always kept subordinate to the lower, and this requires a regular and constant attention.

The Filbert.—The remarks on the quince may be applied with equal propriety to the filbert, as regards form. The *bush* branched from the ground, and the *low standards* with two feet stems, are the ordinary forms; but in some of the French gardens it is conducted with great success as a pyramid.

These are the principal trees of which it is necessary to speak in regard to form. Other species will be referred to under the head of pruning. Having now pointed out the most eligible forms for garden trees, and their respective advantages, planters will be able to make a choice adapted to their tastes and circumstances. Those who do not employ a professional gardener, and who have but a small portion of spare time to devote to their garden, should by all means adopt such forms for their trees as require the least skill and labor, provided always that it be appropriate to the size of the garden, and consistent with good management.

The next point to be considered is,

The Age of the Trees.—This will depend very much on circumstances. For pyramidal trees it is yet difficult, almost impossible, to obtain in the nurseries specimens of more than one year's growth that are suitable. The yearlings are never sufficiently cut back, nor the branches of the second and third year so managed as to have the requisite proportion of length and vigor to fit them for being moulded, with any ordinary treatment, into a perfectly pyramidal form. If suitable trees cannot be found of two or three years from the bud or graft, vigorous yearlings, worked at the ground, should be chosen, as they are in a condition to take easily any required form; and though fruit may not be so soon obtained from them, yet they will in the end be much more satisfactory; for, unless a right beginning be made in the training of a tree in any form more or less artificial, no art can afterwards completely correct the errors. If we take a two or three year old tree, managed in the nursery, as usual, with a naked trunk two to two and a half feet from the ground, and a branching head, or what is nearly as bad, a few weak side branches below, overrun with strong ones above, the most severe process will be necessary, in order to produce lateral branches in the proper place; and thus, as much time will be lost as would bring forward a yearling, and the tree will not be so perfectly formed, so healthy, nor in any respect so satisfactory. The general impatience that exists in regard to the growth and bearing of trees is the great cause of this defective character when taken from the nursery. The nurseryman is averse to cutting back his trees, as they lose a year in height, and planters or purchasers are not generally discriminating enough to be willing to pay him a proportionate price. He finds tall trees more attractive. When planters do get these trees, they cannot be persuaded to cut them down; they wish to obtain fruit as soon as possible, and therefore the tree is allowed to proceed in the defective form it assumed at the nursery.

For *Dwarfs* and *Dwarf Standards*, it is less difficult to obtain the right sort of trees, for this is the form that nursery trees that have not been cut back, ordinarily assume. Those, therefore, who prefer such trees can always be supplied with them well advanced, even in a bearing state if so desired. As in the pyramid, however, persons who intend to make models of their trees, will do well to procure yearlings worked at the surface of the ground, for on them heads or lateral branches can be formed without any difficulty at any desired point between the collar and terminal bud. Another consideration is worthy of note on this point. There is a much greater risk in removing three or four year old trees than yearlings, and they are more difficult and expensive to pack and transport. The yearling is easily removed and easily transported, and its growth is comparatively unaffected by the change. The gardeners most famous for their handsome, well managed fruit trees, invariably select yearling trees, that is, trees that have made one year's growth from the bud or graft.

Selection of Varieties.—The selection of varieties of fruits for a fruit garden should be made in view of all the circumstances that can affect their usefulness. They should be adapted to the soil, and more particularly to the climate. It is well known that in every section of the country, certain varieties seem to succeed remarkably well, whilst others, of the greatest excellence elsewhere, entirely fail. Our country is so extensive and embraces such a variety of climate that it is impossible that the same varieties should succeed equally well in all parts; and planters should consider this well. Those who have had no experience in cultivation, nor a proper opportunity for acquiring knowledge on this point, should consult oth ers. Any intelligent nurseryman who has a correspondence with all parts of the country, and is thoroughly alive to all the branches of his profession, and the results of experience, can aid planters greatly in making appropriate selections. It is true that the amount of knowledge collected on this head is yet comparatively small, and quite insufficient for a general guide, but it is every day accumulating, and what there may be, is well worthy of attention. The experience of fruit growers, as elicited at recent pomological conventions, has brought to light a multitude of highly important facts, bearing on this very point. These will be more particularly noted when we come to the *description* of fruits.

Varieties should be adapted in their growth to the form they are to be grown in, and to the extent of the Garden.—For pyramidal trees, varieties should be chosen whose habits of growth are regular or slightly spreading, the branches assuming more of the horizontal than the upright, and those disposed to branch low down should be preferred to those of an opposite habit. Where the garden is small, moderate or slow growers should be preferred to rapid and vigorous growers. They should also be well adapted to the stock on which they are worked. This is a very important point, but one on which only a few persons in this country have yet acquired any considerable amount of actual experience. Still, many important facts have been gathered, and it becomes every planter to avail himself of them. If he plants pears on quince stocks, for instance, it is important to know that certain varieties are much better on that stock than they are on the pear; and that others fail, and are worthless on it.

The varieties should be adapted to the wants and wishes of the planter.—Those who plant fruit gardens have not all the same objects in view. One man plants his garden for profit, to supply his family with good fruits. This is his main purpose. He should, therefore, select the very best varieties, considering not the *quality* alone, but their productiveness and other useful properties. Such a person has no desire for a large collection, but looks merely for an assortment that will yield a succession of ripe fruits during the season. Another who regards the mere value of the fruit *less* than amusement, recreation, and experiment, will make his collection as varied as possible. Where any particular class of fruits can be had very cheap in market, it should be planted sparingly in the garden, so that such as may be scarce or dear can be grown in larger quantities. It is only by taking all these into account, that planters can hope to make their fruit garden answer their particular views and purposes.

The planting of a fruit garden should be considered as of equal importance, as far as the doing of it well is concerned, with the building of a dwelling. This is constructed with a view to the convenience of the family, and is, therefore, in all its parts, supposed to be adapted to their wants and mode of living. The fruit garden is intended, also, to promote the comfort and convenience of the family, and should, like the dwelling, in all respects be as nearly as possible adapted to their wants and circumstances. Having now treated of the soil, enclosures, trellises, walks, arrangement, selection of trees and varieties, we proceed to the taking up of the trees and planting.

Taking up the Trees.—This has already been described under the head of nursery operations, to which the reader is referred.

Planting has been described under the head of *planting the orchard;* and the operation being the same in both cases, it need not be repeated.

The arrangement of the trees, however, is different, and this point requires a special notice.

1st. *In regard to position.*—Each class of trees, such as pears, apples, cherries, etc., should be planted together in the same rows or division, and if any difference exist on the soil, each should be planted in that best adapted to it. Thus, plums should have that most inclined to clay; pears and apples, the deepest and richest; cherries, peaches, apricots, etc., the dryest and lightest.

Where the garden is large, the pyramids should be in one compartment, the dwarf standards in another, and the dwarf bushes in another; but where it is necessary to economize and fill the ground to the best advantage, the dwarf bushes may alternate advantageously with the pyramids or dwarf standards, and this especially along the walk borders. Varieties, too, of the same, or similar habits of growth, should, if possible, be together. The espalier trees should be placed so that the earliest blossoming kinds, such as the apricots, will be most secure from the influence of spring frosts where these prevail. The trellis facing the north will be the best for this purpose; but where it is intended to protect them, the aspect is of little account. In the north aspect, fruits are very much retarded in their ripening; and this circumstance may be turned to a good account to prolong the season of some late cherries, currants, etc. We have seen fine Morellos in perfection on a north wall here, in the month of September.

The distance at which trees should be planted in the garden.—This will not be the same in all cases; for in a large garden it is not necessary to plant so close as in a very small one, and in a very rich and deep soil, a greater distance will be required than in a dry and light soil. There is also a great difference in the growth of varieties. Some might be planted at six feet apart, and have as much space in proportion as others would at eight. This shows that no rule, as regards distance, can be observed in all cases, and this particularly in small gardens, where advantage should be taken of every circumstance. In large gardens an uniform distance may be adopted, even if some space be sacrificed. The following distances may serve as a general guide, and may be increased or diminished according to circumstances:

DISTANCES IN THE OPEN GROUND.

Apples.—Pyramids on free stock, ten feet apart; do., on Doucain, eight feet apart; do., dwarf standards on Doucain, eight feet apart; do., dwarf bushes on paradise, five to six feet apart.

Pears.—Pyramids on free stocks, ten to twelve feet apart; do., on quince, six feet apart; do., dwarf standards on quince, six to eight feet apart.

Plums.—Dwarf standards, eight to ten feet apart; do., pyramids, eight to ten feet apart.

Cherries.—Pyramids, hearts, and bigarreaus, eight to ten feet apart; do., dukes and morellos, six to eight feet apart; do., dwarf bushes of morellos, five to six feet apart.

Apricots.—Dwarf standard on plum, eight to ten feet apart; do., pyramids, six to eight feet apart.

Peaches.—Low standards on peach, ten to twelve feet apart; do., on plum, eight to ten feet.

Nectarines.—Same as peaches.

Quinces.—Pyramids or bushes, six to eight feet apart.

Filberts, do., six to eight feet apart.

Gooseberries and Currants, four to five feet apart.

Raspberries, two to three feet apart.

Mr. Rivers gives the following distances in his "Miniature Fruit Garden."

THE FRUIT GARDEN.

Pyramidal Pear Trees, on quince stocks, root pruned for small gardens, four feet apart. The same, in larger gardens, not root pruned, six feet apart.

Pyramidal Pear Trees, on the pear stock, root pruned, six feet apart. The same roots, not pruned, eight to ten feet—the latter if the soil be very rich.

Horizontal Espalier Pear Trees, on the quince stock for rails or walls, fifteen feet apart.

Upright Espaliers, on the quince stock for rails or walls, four to six feet apart.

Horizontal Espaliers, on the pear stock for rails or walls, twenty to twenty-four feet apart.

Pyramidal Plum Trees, six feet apart.

Espalier Plum Trees, twenty feet apart.

Pyramidal Apple Trees, on the paradise stock, root-pruned for small gardens, four feet apart. The same roots not pruned, six feet apart.

Espalier Apple Trees, on the paradise stock, fifteen feet apart. The same, on the crab stock, twenty to twenty-four feet apart.

Peaches and Nectarines for walls, twenty feet apart.

Apricots for walls, twenty-four feet apart.

Cherries, as bushes on the mahaleb stock, roots pruned for small gardens, four feet apart. The same, roots not pruned, six feet apart.

Espalier Cherry Trees, on the mahaleb, for rails or walls, twelve to fifteen feet apart.

DISTANCE FOR ESPALIER TREES ON WALLS OR TRELLISES.

The distances between espalier trees must be regulated not only by the growth of the species and variety, but by the height of the wall or trellis. If these be low, a greater length, of course, will be necessary than if high; for every tree must have a certain extent of surface to be spread upon. Hence, if a trellis be only eight feet high, nearly double the length, and, consequently, double the distance between the trees will be required that would be on a trellis fifteen or sixteen feet high. As a general thing, *peaches*, *apricots*, or *nectarines*, on walls or trellises eight or ten feet high, should be fifteen to twenty feet apart, if on free stocks, and twelve to fifteen if dwarfed on the plum. *Cherries*, ten to twelve feet. Our *native grapes*, *Isabella*, *Catawba*, etc., at least thirty feet apart, on an eight feet high trellis, as their rapid growth covers a great space in a short time. Foreign varieties will not require half this; indeed, the better way is, to keep these trained to simple stakes, and planted in the border, where their out-door culture is attempted. In this way they are easily laid down and protected.

CHAPTER II.

PRUNING APPLIED TO THE DIFFERENT SPECIES OF FRUIT TREES UNDER DIFFERENT FORMS.

SECTION 1.—PRUNING THE APPLE AND THE PEAR.

THESE two trees belong to the same natural order, *pomaceæ*, and to the same genus *pyrus;* their habits of growth and bearing are similar, and they may therefore be treated as regards their pruning, under the same head.

If we take for example a shoot of last season (fig. 6), we find it in the spring, before vegetation commences, furnished on all its length with wood buds; when growth commences, the terminal bud, and probably two or three of the others nearest to it, produce shoots, the others towards the middle produce small shoots that are in subsequent years transformed into fruit branches (like fig. 10). Some do not push at all, but are converted into fruit buds (as in figs. 7 and 8), whilst those at the base generally remain dormant, until excited into growth by close pruning. All the buds on these trees have small inconspicuous buds at their base, which are capable of producing shoots when the principal bud is destroyed or injured, and these buds render the fruit spurs so enduring. In young trees the fruit buds are many years in process of formation, and in bearing trees three to four years, according to circumstances. When the trees are not subjected to pruning, the result of the mode of growth described is, that the terminal buds grow and form one section upon another, leaving the lower parts mainly destitute of bearing wood, unless it be an occasional spur, the sap always tending to the points.

1st. *Standards.*—The management of this form of trees has been fully treated of in all our works on fruit culture, and in all the agricultural and horticultural journals, so that now it is pretty well understood, and especially by those who give considerable attention to the subject of fruit trees; it will not be necessary therefore to enter upon much detail in regard to it.

A standard apple or pear tree for the orchard, when taken from the nursery to be finally planted out, we will suppose to have a straight, stout trunk, four to six feet in height, as the case may be, and a head composed of a certain number of shoots or branches, but generally shoots of one year's growth. At the time of planting, three or four of these shoots should be selected to form the main branches, or frame-work, on which to build the whole head, and the remainder cut clean out; those reserved should be cut back full one-half, and from the shoots produced on these at and below the cut, two of the strongest are selected each on opposite sides, and the others are rubbed off while they are soft. In selecting these shoots, care must be taken to have them equally distant from one another, and pointing in such directions as not to cross or interfere.

During the first season these young shoots must be watched and kept in a regular state of vigor. If any threaten to become too vigorous, they must be pinched and checked at once, so that perfect uniformity be preserved. This is the time to secure a well formed and nicely balanced head. A very slight circumstance some-

THE APPLE AND THE PEAR. 205

times throws the growth into one side or one branch of a young tree, and produces a deformity from which it never recovers. The trunk must be kept clear of all shoots, by rubbing off such as appear at the earliest possible moment, when it can be done without the use of a knife. Supposing we commenced the head with three branches at time of planting, there will be at the end of the first season, six.

The attention required after this will be to maintain an uniform growth among these six branches, and their members and divisions, and to prevent the growth of shoots in the centre. The leading defect in all our orchard trees is *too much wood*, the heads are kept so dense with small shoots that the sun and air are in a great measure excluded, and the fruit on the outside of the tree only is marketable or fit for use. The head should be kept open, rather in the form of a vase, so that the wood, leaves, blossoms and fruit may all, on every part, enjoy the full benefit of the sun and air, without which they cannot perform their functions, or maintain maturity and perfection.

Too many people imagine that trees can take care of themselves, as trees in the forest, on the ground that nature preserves a balance in all her works; but it should be borne in mind that a fruit tree is not exactly a natural production. It is far removed from the natural state by culture, and the farther it is removed, that is, the more its nature is refined and improved, the more care it requires. Fig. 103 represents a young standard pear tree, stem four feet high, and the head twice cut back, as at the letters *a* and *b*.

FIG. 103

A young standard pear tree, trunk 4 feet high, head formed on three main branches, twice pruned as at *a* and *b*

Pinching.—If this be properly attended to, very little knife pruning will be necessary, except to shorten the leading shoots, because as soon as a superfluous or misplaced shoot appears, it is rubbed off, and when one becomes too vigorous, it is pinched and checked; the great advantage of pinching is, that 1st., It economizes the sap of the tree. That which would be expended on superfluous shoots is turned to the benefit of the parts reserved, and thus the growth is greatly promoted.

2d. All wounds necessarily inflicted, where knife pruning is depended on, are completely avoided. These facts should be remembered. Standard apples and pears are not generally pruned with a view to hastening their bearing, but are allowed to arrive at that state in their natural way. In the case of tardy bearing sorts, however, it *may* be desirable to apply artificial means, and these will be pointed out in treating of dwarfs and pyramids hereafter.

Dwarf Standards.—These are similar to standards, except that the trunks are low, not over two or three feet in height, and the head is retained in a smaller space. Their management is always much easier when the stocks are such as to dwarf or restrain the growth. Thus, apples on the *paradise* or *Doucain*, and pears on the *quince*. The main branches or frame-work of the head, are produced by cutting back the three or four branches that form the head of the tree as it comes from the nursery, in the same manner as recommended for standards.

The *first season*, all superfluous productions are rubbed off, and a balance maintained among the shoots by pinching.

The *second year*, in the winter or spring, the shoots of last season are shortened, say one half, as a general thing. This induces the development of the buds on their whole parts. The cut is made at a good, plump bud, capable of producing a vigorous shoot; and this is selected to prolong the branch. If one or two secondary branches are needed to fill up a space, those next the leader, if properly situated to fill the space, are chosen, and all below them are pinched when about two or three inches long, in order to check the production of wood where it is not wanted, and to convert them into fruit branches or spurs. The growth of all the main and secondary branches is regulated and balanced by pinching; and if the pinched shoots intended for fruit spurs start again into growth, they must be again pinched.

The *third season* the shoots of the previous year are cut back as before, say to four, five, or six eyes, according to their strength. One shoot is chosen to continue the prolongment of the branch, and the others are pinched in season to convert them into fruit spurs. Thus the tree is conducted from year to year, until it has attained the full size required. In this way the trees commence bearing quite young, and every branch is furnished in all its length with fruit spurs.

Pyramids.—Under the head of "the selection of trees," it has been recommended to obtain thrifty yearling trees in preference to older ones not properly managed. We will, therefore, begin with the yearling tree, and although the management of this the first year after cutting back has been given in the nursery, it may be well to repeat it here, to save the reader the trouble of referring back.

Objects of cutting back.—The object in doing this is to produce branches near the stock that will form the base of the future pyramid. If left entire, the tendency of the sap to the extremities would produce shoots there only, leaving a naked space entirely inconsistent with the form in view. We, therefore, reduce the stem to such an extent, that but a small number of buds is left on it, and the sap acting on these with great force causes their development.

How far to cut back.—It is obvious that this must depend on the character of the subject. In yearling plants, both of the pear and apple, there is presented a great difference in different varieties. Some invariably produce lateral branches the first season. The buds are so perfectly developed, that when the second growth takes place in midsummer, they break and form branches, in some cases as much as a foot long, and in others only a few inches. Then among the varieties which do not thus produce side branches in the second growth, there is a great difference in the plumpness and prominence of the buds. In some they are larger, and stand out boldly from the wood on the whole length of the stem, apparently ready to push under the least excitement. In others they are small, lie flat to the wood, and have every appearance of being difficult to excite into growth, and especially those towards the base. It should always be borne in mind that it is better to cut *too low* than *not low enough*. The difficulty of cutting too low is, that the shoots produced are nearly all of equal length, and a certain number of them require to be checked to give each one its proper dimensions. The difficulty of not cutting low enough is, that where we should have branches at the base we have none, or, if any, they are smaller, instead of larger, than those above them. The remedy in this case is more difficult than the other. The vigorous shoots at the summit must be checked, and even the leading shoot, in order to throw back the sap into the lower parts to act upon the buds there. The error which produces such a difficulty, is very common, as we know by experience, amongst persons not familiar with the growth of young trees or the development of the buds on their stems. It must be laid down as a general rule, that *the more feeble the plant, and the smaller and the more imperfectly developed the buds, the lower it is necessary to cut.*

The condition of the roots, too, must be taken into account; for where the roots are weak, broken, or injured, and consequently unfit to yield to the stem any considerable amount of nutriment, the buds will break with less force, and a more severe retrenchment will be necessary. All these circumstances must be considered. For example, we will take a young pear tree of one year's growth from the bud, without branches (fig. 104), which we will suppose to be four feet, which is the ordinary average height of yearlings. If the buds are full and prominent on it, we cut to a good bud at twenty inches from the stock; but if the buds are less prominent, cut to fifteen or eighteen inches, and if *very* feeble, with small buds, cut to within twelve inches, or five or six buds of the stock. If the roots have been injured much, and the stem somewhat dried or shrivelled, it should be cut to within three or four buds of the base. These different cases are mentioned because it frequently happens that persons who live at a great distance from nurseries, find their trees frequently, on their arrival, in the condition described, and it is necessary that a course of treatment for them should be indicated. The bud cut to, should, if possible, be one of the best on the stem, and be on the side of the tree opposite that in which the bud was inserted, so as to continue the stem in a straight line.

Fig. 104.

A yearling pear tree without branches. The cross-line indicates the first pruning or cutting back.

It is a great advantage to have a tree well established in the ground, before cutting it back to produce the first branches to form the pyramid; because, in that condition, it is capable of producing vigorous shoots the first season.

It is on this account that a young tree, cut back in the nursery, presents a much more perfect form at the end of the second year, than those that have been transplanted. Some of the French cultivators advise to defer the cutting back for the formation of the permanent branches, till the plant has stood one year after transplanting; but the course is attended with many difficulties, and on the whole it is better to cut back when the tree is planted, even if we obtain but a moderate growth, for the older the buds are on the lower parts of the tree, the more obstinate and unmanageable they are.

Pruning the Branched Yearling.—Among trees of this kind, some have branches a foot or more in length, while in others they resemble short, stiff spurs, two to four inches long. These two characters require different modes of treatment. Where there are branches of sufficient force and properly situated to form the first series of main branches, they must be treated in the same manner as though the tree were two years old. The strongest and best situated are selected and pruned to within four to six inches of their base, according to their vigor and position; the lowest should be not more than six inches from the stock. The small, feeble, superfluous ones are entirely removed; the leading shoot, which, in such cases, is short and provided with plump buds, does not require a heavy shortening; in most cases one half will be quite sufficient. Fig. 105 represents a tree of this kind; the cross-lines indicate the cuts. Where the lateral branches are short and spur-like, they will require very careful treatment; the strongest and best placed are reserved. If the lower ones have good terminal buds, they are left entire; those above them are shortened, the

Fig. 105.
Yearling pear tree with branches, the pruning indicated by the spaces.

lower to three, the next above to two, and the uppermost, next the leading shoot, to one bud. This will give their productions a proper relative degree of vigor. The leader is cut back further than in the well branched subject, because it is presumed the buds are less excitable. As a general thing, within four to six buds of the highest lateral, or one half of its length.

There is another class of trees necessary to be noticed here, because they are very common—*two year old nursery trees that have not been properly treated*. Fig. 106 represents a tree of this kind. A few inches only of the top were taken off at the commencement of the second year's growth, and after that it was left to itself. Branches, therefore, were produced only at the top, leaving a vacant space of two feet, the very part that should have produced the first set of main branches. The best disposition to make of such a tree would be to conduct it in the form of a dwarf standard, which it really is at present; but it happens that in some cases it is desired to convert them into pyramids, and therefore it is essential that the proper means be pointed out. Two year old trees, like yearlings, differ materially in the character of the buds on the lower part of the stem. On some, these are quite prominent, so much so as to appear to have made some advance towards development, while in others they are quite flat and dormant. It is obvious

Fig. 106.
A two year old pear tree, not cut back far enough the first season; the second pruning, to produce branches below, is indicated by the cross line.

that trees in the first condition will not require that severe retrenchment on the head to produce branches below, as the last. In this case it will generally be sufficient, and especially if the space between the stock and first branches does not exceed two feet, to cut back the leader to three buds, and the lateral branches below it to one bud; but when the buds are small and backward, or when the branchless space is over two feet in length, the two year old wood must be cut back to within eighteen inches to two feet of the base. We find that in the case of imported trees, or those carried a great distance, and more or less injured, nothing short of this severe cutting can ensure branches low enough to form a pyramidal tree. It seems a great pity to cut back a tree in this manner, and lose a year or two of its growth and bearing, but it is absolutely necessary when the pyramidal form is wanted. There is still another class of trees that we sometimes see sent out from the nurseries. These are two or three years old; have been cut back, and are pretty well furnished, in all their length, with lateral branches; but from the want of proper care, those on the upper parts have acquired greater vigor than those below, presenting the tree in a situation just the reverse, in this respect, of what it ought to be. In pruning this subject at the time of planting, the lower branches must either be shortened very slightly in order to get a strong bud for a leader, or they must be left entire, while those above will be cut close; where we want the longest and strongest branches, there we leave the most wood.

The most important pruning performed upon a tree is the *first one*, for it is this which makes all future management easy and successful, or difficult and unsatisfactory. This is the reason why it has appeared necessary to treat of it so minutely. Having encountered all the difficulties that others are likely to encounter, and having described them and pointed out the means by which they are to be overcome, it is believed that the matter has been made so plain, that any man of ordinary intelligence, and possessing the slightest knowledge of tree culture, can take his knife and prepare his trees in such a manner as to give him a most reasonable hope of attaining his ends. We now proceed to the

Summer management of trees thus cut back.—We will first consider the case of the yearling without branches. If it has been cut low enough, as directed, all the buds below the cut will push. As a space of six inches should be kept clear between the ground and the first or lowest tier of branches, such shoots as may appear on that part will be rubbed off at once. Of the remaining ones, a certain number, three to six, according to the length of the stem, will be reserved. These must be the strongest, and properly situated on the stem, within eight to ten inches of space between each branch, and that immediately above it, and regularly placed on all sides of the stem. Some recommend leaving on all the shoots that are produced the first season; but in certain cases this would be bad practice, for if the buds be very close, the shoots would be so numerous that the strength of them all would be impaired, and much pruning would be required the next season. The better way is to select such as are wanted, and rub off the others; the sap which they would have appropriated will be turned to the account of the permanent branches, and increase their vigor. The leading shoot must be directed in a straight line; in some cases a support may be necessary. If the branches immediately below it are so vigorous as to interfere with its growth, they must be checked by pinching. In some cases it may be necessary to do this when they are an inch or two in length. It sometimes occurs that the bud cut to is injured by the weather, close cutting, or some other cause, and pushes so feebly that the laterals below it having more vigor take the lead. This must be prevented in time. A proper relative degree of vigor must be maintained among all the branches, by checking when necessary the most vigorous.

The first summer's treatment of the branched yearling (fig. 105.) will consist in maintaining a uniform growth among the lateral branches, and in the case of the leading shoot, as already described. Some lateral shoots will be produced on the branches, and these must all be pinched at an inch or two, as it is yet too soon to allow of the formation of secondary branches. The summer treatment of fig. 106, the two year old tree, will be conducted on the same principles. The encouragement of the leading shoot will require special attention to secure it in an upright position, as, in many cases, where two year old wood is cut back, the leading shoots assume a horizontal or curved direction.

The second pruning.—We have now a tree composed of two sections: the first is the two-year-old part, furnished with lateral branches; and the second, the leading shoot produced last season. (Fig. 107.) In pruning it, our object will be to establish a new section of branches on the leader, to continue the prolongment of the lower branches, and to induce the formation of fruit spurs towards their base. To accomplish these ends, we shorten the leader or stem, on the same principle in relation to its character, as already directed for the yearling trees, from one-half to two-thirds its length, and sometimes more. Every bud between the one we cut to and the base of the shoot, should push; and the bud to produce the leader should be large, perfectly formed, and *opposite* the cut of the previous year. The lateral branches on the first section are shortened according to their vigor, always remembering that the lowest must be the longest, to carry out the pyramidal

Fig. 107.
A two year old pear tree, having made one year's growth after the first pruning.

form. They should also be cut back sufficiently to insure the growth of all the buds on them. This point requires considerable care, for if not cut back enough, the interior of the trees becomes naked, instead of being supplied with shoots for bearing spurs; and if cut back too far, the shoots will be too vigorous and difficult to control. The appearance of the buds, and the habits of the variety, will be a sufficient guide if properly studied.

Treatment of the growing shoots.—When the buds have all started and made a growth of an inch or two, their force and forwardness will indicate the uses to be made of them. Each of the main branches of the first section may be considered as a stem; its leader will require the same treatment to favor its extension. At this time a secondary branch may be required to fill up the space which widens as the branches extend. If so, a shoot is selected for this purpose, and all the others on the same branch are checked at two inches, and converted into fruit branches. All the laterals are treated in this way. The second section, now in process of formation, must be managed as directed for the first section. During the first season, the requisite number of shoots is preserved, and the superfluous ones removed early. The leader is maintained erect; and the laterals immediately below it, being always inclined to vigorous growth, must be checked to keep them in a proper condition relative to the leader and the branches below them. The leading shoot must always maintain its pre-eminence. It often happens that the lateral shoots of the main branches that have been pinched will start and grow again. In such cases another pinching must be performed within an inch of the previous one. As a general thing, this will be sufficient; but if not, a *third* must be given in the same way; for if they be allowed to extend into wood branches they will require knife pruning, and create confusion among all parts of the tree. A very general error in conducting trees of this kind, and indeed all others, is to allow the branches to be too close to each other, so that when they come to bear, the wood, foliage, and fruit, on the interior, are so excluded from the air and light that they all suffer. The fruit is imperfect, and the spurs become feeble and gradually perish. The tree has now two branched sections, each from twelve inches to two feet, as the case may be, and with four to six branches on each; the leading shoot is from one to three feet in length.

The average height of three year old trees, on the quince in our grounds, transplanted at one year old, and twice pruned, is five to six feet. A few very vigorous growing varieties, that throw up a leader every season three to four feet in length, are seven to eight feet; but these are comparatively few in number.

Third pruning.—This is done on precisely the same principles laid down for the second. The leader of the stems is cut back in proportion to its vigor, the lateral branches are also shortened in the same manner. It must always be kept in mind that the

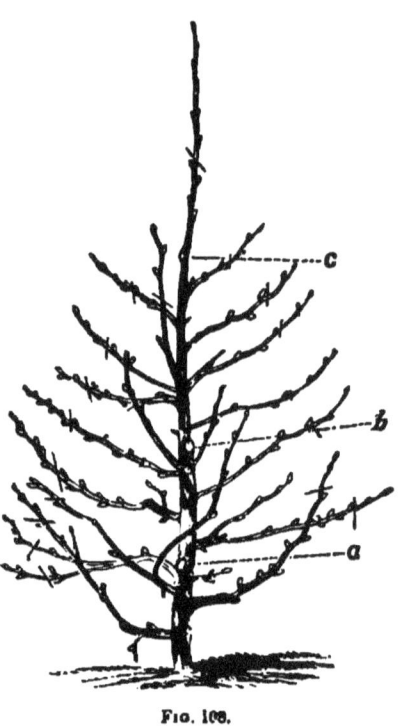

Fig. 108.

A pear tree four years old, three times pruned, having three branch sections, *a, b, c,* seven feet high, and furnished in the lower parts with fruit spurs. The cross lines indicate the fourth pruning.

lowest branches must be longest, and when it happens that they do not take their due proportion of vigor, as compared with those above them, and if pinching has not been duly attended to the past summer, to maintain regularity, the weaker must now be favored with a *long* pruning.

Fig. 108 represents a four-year old pear tree of the white Doyenne, three times pruned, *a, b, c*.

It has been remarked, that the habits of growth and bearing of the pear and apple are similar, but it should be noted, that in treating them as pyramids, the apple tree is more liable to lose its vigor at the top, and therefore it is necessary to keep an eye to this point in their management. From what has been said with reference to an equal distribution of the sap, the remedy for this difficulty will be obvious, viz., to reduce the vigor of the lower parts by pinching, shortening, and heavy crops, and to favor the upper part by long pruning and thinning, or wholly removing the fruits.

Management of the fruit branches.—About the sixth to the eighth year, from the first pruning of the tree, it will have attained nearly as great dimensions as in many cases will be desirable, and be well furnished with fruit branches.

After this period, the object of the pruning will be to prevent the extension of the tree, and maintain the fruit bearing parts in a healthy and productive state. Without proper care they will be liable to suffer from bearing too much, or from the growth of young wood on the extremities of young branches.

To diminish the growth, and favor the fruit branches, the young shoots must be pruned shorter than before, in order to turn the sap more to the benefit of the fruits, and when the fruit spurs become too numerous, so as to be too near one another, and produce more fruit than the tree can sustain with safety, a portion of them must be pruned off. The lower parts always experience this difficulty first, the sap circulating more slowly there than in the summit. Fruit spurs of the pear and apple, if well managed, continue in a vigorous bearing state for a great many years. To renew and prolong their vigor, the older parts must, from time to time, be cut away, and new productions created at their base to take their place.

Pruning and management of the Apple as a dwarf on the paradise stock.

Nothing is more simple than the treatment of these little bushes.

They should have short stems, six to eight inches from the ground, and the head should not be allowed to exceed three to three and a half feet in height, because the roots are very small, and do not take such a firm hold of the ground as to admit of a head that would offer much obstacle to the wind. The branches should be evenly distributed around the head, open in the centre, in the form of a vase, and be furnished in all their parts with bearing spurs.

These are the points to aim at in commencing the formation of these trees. The proceedings are as follows:

1st. *Pruning.*—We will suppose that the subject is a yearling bud or graft, a single shoot eighteen to twenty inches in height. In this case, the stem is cut back to the point where it is intended to form the head, six to ten inches, as the case may be, from the stock. Below this, most of the buds will start and form shoots, from which we select three or four of the strongest and best situated, equally distant, if possible, around the stem, and rub or pinch off all the others. The growth of the branches thus selected for a head, is encouraged during the first season, by keeping down all other productions that may appear.

2d. *Pruning.*—The tree has now three or four branches destined to be the basis of the frame-work of the head. These branches are cut back full one half their length, according as the buds in the variety are easily excited or not, the object being to induce all the buds below the cut to push. After growth has commenced, and an inch or two of new wood been made, the shoot from the bud cut to, will be chosen as a leader to continue the extension of the branch; and if secondary branches be wanted, they will be chosen from those best situated, to fill up the existing vacancies. All the other shoots are pinched when two or three inches long, to convert them into fruit spurs, and to prevent their interfering with the growth of the wood branches. If one pinching is not sufficient, another must be given in the same way as recommended for pyramidal trees. Indeed, the whole process, as far as it goes, is the same; but the same efforts are not necessary to maintain an equal distribution of the sap, for the tree is so low, and the form so natural, that no branch is more favorably situated than another; and hence they are easily kept in an uniform state of vigor. The branches of irregular-growing sorts will require to be secured by stakes in their proper places for a year or two at first, until they have assumed a permanent position.

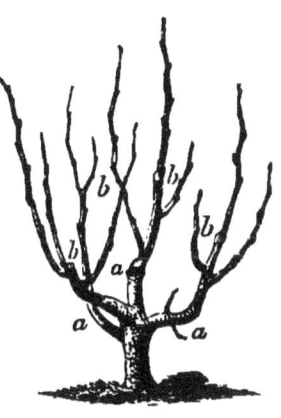

Fig. 103.

Dwarf apple tree, four years old, stem ten inches high, head composed of four main branches and several secondary branches: pruned three times as at *a*, *b*, now in a bearing state.

The third, and all subsequent prunings, will be conducted on the same principles as the first and second, already described, until the tree has attained its full size. Fig. 109 represents a dwarf apple tree, four years old, three times pruned—the two last prunings are indicated by the letters *a* and *b*.

Management of the Bearing Tree.—In most cases the apple on the paradise is disposed to excessive fruitfulness, and unless the fruit branches be occasionally thinned and shortened, in order to reduce the number of bearing buds, and to produce new wood, the trees become enfeebled. Bad management of this kind has promulgated the belief that the apple on the paradise is exceedingly short-lived; but the fact that plantations exists in the most perfect vigor at the end of twelve to fifteen years after planting, shows that by proper treatment their existence is not so fleeting. The spurs must be managed in a manner similar to that described in treating of pyramids, to renew them, and the slender fruit branches must be shortened. This, in addition to the manuring to be hereafter described, constitutes the substance of their management.

The Pruning and Management of the Apple and Pear as espaliers.—In the cool, moist climate of England, this is a popular and advantageous method of training apples and pears. The specimens of this kind in public and private gardens there, are admirable in their way, and illustrate the skill and handiwork of the English gardener very favorably. But our climate is not suitable as a general thing for espaliers; the branches are so exposed to the rays of our powerful sun, that the sap is impeded in its circulation, and the fruits fall. It is, therefore, unnecessary to enter into any detail respecting this mode of training; but there may be situations where such a system may succeed, and especially in the north. The best espalier form for the apple and pear, is that of the *horizontal*, that is, an upright central stem, with horizontal arms or branches at equal distances on both sides (fig. 110). The production of this tree depends in the main on the same principles as the pyramid, and does not require illustration. The young tree is cut back to within six inches of the ground. From the shoots produced be-

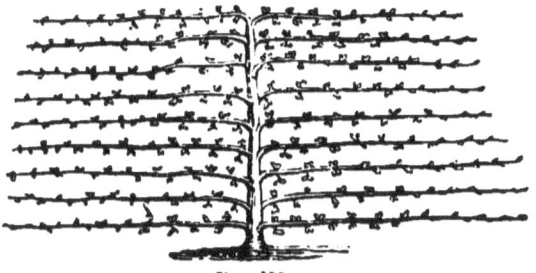

Fig. 110.
Pear tree trained horizontally.

low that point, three are selected, the upper one to form the upright leader or stem, and two lateral or side ones to form the two first arms. The first season these shoots are allowed to grow upright and are kept in equal vigor. At the commencement of the second season, they are all cut back far enough, say one third to one half their length, or even more in some cases, to ensure the growth of all the buds.

The upper shoot on each is selected for a leader, and the others are pinched at two inches or less. After the pruning, the arms are brought down half way to a horizontal position, and towards the latter end of the season, wholly. An uniformity of growth among all the parts is maintained according to the means and principles already laid down, and year after year the tree is thus treated until the requisite height and number of horizontal arms or branches be obtained. In the case of very vigorous growing sorts the leader may be stopped in June, and thus a second pair of arms be produced in one year. The upright leader and the branches are treated in a similar manner—a difference in vigor always requiring a corresponding difference in treatment. For espaliers, the apple should be on the *paradise* or *Doucain*, and the pear on the *quince*, because these stocks all diminish the vigor of wood growth, which is often the chief difficulty in man aging trained trees.

The aspect for these trees should never be due south. A railing to train such trees on, is made of upright posts sunk in the ground, and connected with cross bars, at eight to twelve inches apart, upon which the arms of the espalier are fastened with willow or bass matting. Mr. Rivers, in his " Miniature Fruit Garden," exhibits a system of growing pears in espalier, in the form of pyramids, as adopted by himself. I saw these trees when in England, in 1849, and although it appeared a very ingenious and economical arrangement, admitting a great number of varieties in a small space, and besides very well adapted to an English climate, yet it did not appear to offer any advantages that would warrant its recommendation in this country, unless under rare circumstances in the most northern sections. Whoever will study attentively the means described for conducting a pyramid, can succeed fully in training the espaliers or wall pyramid.

SECTION 2.—PRUNING AND TRAINING THE QUINCE.

As ordinarily grown, the quince is the most neglected, and consequently, the most unsightly, deformed tree to be found in the orchard or garden, and yet, when well treated, it is really, both when in blossom and in fruit, one of the most beautiful of all our fruit trees. Its fruit is more esteemed, and more generally used in this than in any other country. It is naturally a crooked or spreading bush, and without some attention to pruning and training when young, it assumes an irregular form, branching near the ground, and quite destitute of bearing wood on all its lower and interior parts. It is in this neglected form we most generally find it. To make a regular and handsome

THE QUINCE.

little tree, we have only, in the first place, to rear a straight and stout trunk about two or three feet high.

If the plants be weak or crooked when planted, they should be cut low down to obtain a stout and straight stem. The young shoot should be kept tied up to a stake to prevent it from straggling.

The second year, if the growth has been vigorous, and low trees are desired, the head may be commenced. But if a stem three or four feet high be desired, it should be at least one inch in diameter, and another year's growth may be necessary.

The head is formed in the same manner as described for standard and dwarf apples and pears. It should be round, symmetrical, and open, and well furnished on all parts with bearing wood.

The bearing branches or spurs of the quince are small twiggy shoots (*B*, fig. 111), produced on wood at least two years old. These bear two, three, or more fruit buds. These produce shoots two or three inches long (*C*, fig. 111), on the point of which the fruit is borne singly. These spurs have always wood buds as well as fruit buds, and therefore they should be shortened back as to *A*, fig. 111, the spring after they have borne, in order to produce new spurs at the same point.

Fig. 111.

B, fruit branch of the quince. *C*, the shoot produced from the fruit bud. *A*, point at which it should be cut back after bearing.

The French conduct it in beautiful pyramids, on precisely the same principle as the pear and apple; but the leading shoot must be kept fastened to an upright support—a small rod attached to the base—on account of its reclining habit. The medlar is but little cultivated. Its treatment may be exactly similar to that described for the quince, its habits of growth and bearing being similar.

Section 3.—Pruning the Cherry.

The cherry is conducted in any desirable form with as much facility as any of all our hardy fruit trees. The *heart* and *bigarreau* classes are very rapid growers, often attaining the height of six feet the first season from the bud or graft, and in two years forming fine standard trees six to seven feet high, with a few top shoots. They have also large, drooping leaves, and, with few exceptions, stiff, erect, or slightly curved branches.

The *duke* class does not grow so rapidly. The branches are stiff and erect, the leaves smaller as a general thing than the preceding classes, more erect, thicker, and of a deeper, darker green color.

The *morellos* are of a bushy habit, with smaller leaves than any of the preceding classes, and the branches are more slender and closer together. The bark of all is very tough, being composed of several layers of powerful fibres and tissue. It does not yield readily, like that of most other fruit trees, to the expansion or growth of the wood, and this occasions the bursting and exuding of gum in certain localities, especially in the more rapid growing classes. The mode of bearing has already been described under the head of fruit branches, in the beginning of the work. The fruit is produced on wood three years old thus: The shoot of last year's growth, which is furnished now with leaf buds in all its length, will produce at the point, if not shortened, one or more shoots, and all the buds remaining are, during the season, transformed into clusters of fruit buds, and produce fruit the year following. In the centre of these clusters of fruit buds there is always a wood bud, and this grows a little and produces new clusters of fruit buds to replace those that have borne. Some of the morellos produce fruit on two-year-old wood, like the peach, the leaf buds being transformed into fruit buds. During the second growth of the first season of their formation, the fruit bud is very easily distinguished from the leaf bud by its roundness and plumpness.

Pruning the Cherry as a Standard.—In Western New York the cherry succeeds so well, and is so totally exempt from the bursting of the bark, that trees can be grown safely with trunks five or six feet high; but in the West, when this malady prevails, the less there be of a naked trunk the better; for it is the trunk and large branches that are generally so affected. As a standard, the cherry requires very little pruning.

To form a round open head.—We will take for example a young tree two years old, having three or four top branches. These at the time of planting should be cut back to within four or five buds of their base, and when growth has commenced, the requisite number of shoots, say four or five, to form the framework of the head are selected, evenly distributed on all sides, and all the others pinched or rubbed off.

The following season these shoots may again be shortened to produce secondary branches to fill up spaces, and those arising in the centre should be pinched out, for the head must be kept open and accessible to the sun and light. In about three years of such treatment, the head of the tree assumes a permanent form, and thereafter, may be left to itself, except to remove occasionally branches that may cross or interfere with one another. Our standard trees here are in the best possible condition, and have not had a knife on them, except to cut scions for budding or grafting, in seven years.

Pyramidal Headed Standards.—Certain varieties, for instance, *Sparhawk's Honey, Downer's, China Bigarreau, Black Tartarian, Black Heart*, and some others, make fine pyramidal shaped heads without pruning, more than to give the leader its due superiority at the beginning, and to remove afterwards crossing and superfluous branches.

Such varieties as the Yellow Spanish, Black Eagle, Knight's Ey. Black, Elton, and all the spreading sorts, should have round open heads built upon three or four main branches as described.

Pruning the Cherry as a Pyramid.—The same process recommended for the pyramidal training of the pear and apple, may be applied with complete success to the cherry. We have now in our specimen grounds a collection of all the classes trained, according to the method described, and their condition is in every respect satisfactory; they have all given fruit the third year.

In most cases the trees were taken from the nursery rows at the end of their first season's growth from the bud. Some had no side branches, and others had. It is very common for cherries and especially the Dukes and Morellos to form a number of lateral branches the first season. Growth becomes slightly suspended, or at least goes on very slowly in July; during this time the buds on the lower part acquire a sort of maturity, and when a new growth commences they push and form shoots. Cherry trees of this kind are in a good condition for pyramids. We select from these the strongest and best situated to form the lower tier of permanent branches; the lower ones are shortened to four or five buds, and the upper ones to two or three. The leader or stem is cut back to within six, eight, or ten buds of the branches. Those having no branches are cut back to within six or eight buds of the stock. And this is the first pruning.

Treatment during the first Summer after Pruning.—When the young shoots have grown a couple of inches in length, such as are intended for permanent branches are chosen, and the others are pinched in the same manner as recommended for pears and apples. Such as acquire more vigor than is consistent with their position, must be checked. It frequently happens that unless the leader has been cut back close, only three or four shoots will be produced at the extremity, leaving a vacant space below. This can be remedied in most cases by pinching the shoots around the leader when they have grown about an inch. In some cases it may be necessary even to check the leader to force the lower buds into growth. This is a point of considerable importance in conducting a pyramid, and should never be lost sight of.

The Cherry as an Espalier.—Except it be the training of the morello, or some other late varieties, on a north wall to prolong their season of maturity, the cherry is seldom grown as an espalier tree in this country, nor is it to be recommended except in some rare instances. The simplest and probably the best form is that suggested for pears and apples, an upright stem with horizontal branches. To produce this the same means are employed as have been previously described. If the tree has no side branches proper for the first arms, it must be cut back to within six inches of the ground, and from the shoots produced below that, one is selected for the leader, and one on each side for the first horizontal branches; the other shoots are pinched off. At the next pruning, the leader is again shortened to produce another pair of side branches eight or ten inches from the first; the leader is continued in an upright direction, and the side branches are brought half way down in midsummer, and at the following spring pruning they are placed in the horizontal position. The leading shoot of rapid growing sorts may be stopped about the end of June, and this will produce side shoots from which another pair of arms may be taken, and thus gain year in the formation of the tree, or covering the wall or trellis.

For weak growing sorts, the fan form or some modification of it would, perhaps, be more suitable than the horizontal, as it offers less restraint to the circulation of the sap in the branches.

The Cherry as a Dwarf or Bush.—The slow growing sorts, such as the *dukes* and *morellos*, when worked on the mahaleb stock, make very pretty and very easily managed prolific bushes, and by occasional root pruning they may be confined to as small a space as a dwarf apple tree. To produce this form, the young tree is cut back to within five or six buds of its base; and from the shoots produced below that, four or five evenly distributed around the tree are selected for the permanent branches or frame-work of the tree. The others are rubbed off. At the next pruning the branches thus produced are shortened to produce secondary branches; and thus it is treated from year to year until the tree is formed and full grown.

The branches must be kept far enough apart to admit the sun and air freely amongst them. When the tree is five or six years old, if it grows too vigorously, requiring more space than can be given it, the larger roots may be shortened in July or August, or in the winter. This and the pyramid, and the dwarf standard, with stems two feet high, are the most eligible garden forms for the cherry.*

---

* Mr. Rivers states in his " Miniature Fruit Garden," that he has a plant of the late duke cherry ten years old, that never was root-pruned, and yet is a small prolific tree, five feet in height, and the branches the same in

The *dwarf standard* is treated precisely as the dwarf, and differs from it only in having two feet instead of six or eight inches of stem. In pruning and training the cherry, it should always be borne in mind that when large branches are removed, it is liable to suffer from the gum, and, therefore, the regulation of the shoots should be carefully attended to in summer, that amputations of woody parts may be avoided as far as possible. When it is necessary, however, the cut surface heals more rapidly and surely when made in the summer, during the growing season.

SECTION 4.—PRUNING AND MANAGEMENT OF THE PEACH.

The peach is universally regarded as the most delicious fruit of our climate, and ranks in importance for orchard culture next to the apple and the pear. Nowhere in the world is it produced in such quantities, and with so little labor, as in America. An English or French gardener will expend more labor on a single tree, than the majority of our orchardists do upon one hundred. Our favorable climate obviates a multitude of difficulties that have to be contended with in other countries, and renders unnecessary the minute and laborious systems of management which they find it absolutely necessary to pursue.

But this very excellence of our climate has given rise to a most negligent and defective system of cultivation, as is everywhere illustrated in the condition of orchards. The peach, of all other trees, is one that, from its mode of growth and bearing, requires constant pruning to maintain it in a shapely, thrifty, and productive state. The sap tends powerfully to the extremities of the shoots,

diameter. We have in our specimen grounds trees of several dukes and morellos, six years old, on mahaleb stocks, not over four or five feet high, and pictures of fruitfulness.

more so than in any other fruit tree. The buds that do not push and form shoots the first season after their formation, are lost; they cannot, as in most other trees, be excited into growth; and hence it is that the lower parts become so rapidly denuded of young wood, and that trees left to themselves for six or seven years are in a measure worn out and worthless.

The fruit is borne only on wood of the preceding year (see fruit branches), and every part destitute of such wood must be worthless; consequently one of the great objects of pruning is to keep all parts of the tree furnished with a regular and constant succession of annual bearing shoots.

This fact must never be lost sight of.

The case of a single shoot will illustrate the influence of pruning and its necessity. By referring to the fruit branch, it will be seen that it is furnished with a certain number of wood buds and fruit buds. At the base there are always one or two wood buds at least.

Now, if that shoot were not pruned, all the fruit buds on it would probably produce fruit—one, two, or three of the wood buds at the top would make new shoots; these would necessarily be very weak in consequence of the number of fruit below them. At the end of the season there would be a long, vacant space, entirely destitute of a young shoot or a living bud. This is the way that the interior and lower parts of trees become so soon degarnished.

But when that shoot is shortened, we will say one half, the sap is retained in its lower parts, one half of the fruit buds are removed, and the consequence is that large and fine fruits are obtained from those remaining; young vigorous shoots are produced from the lower buds to bear next year, and take the place of those which have already borne. In this way regular uniform crops of

THE PEACH.

large and fine fruit are obtained, and a constant succession of young shoots is kept up.

To form the head of a standard Peach Tree.—We will suppose it the intention to form a standard tree, with a trunk two feet in height, and a round, open, and symmetrical head like fig. 112. We take a yearling tree and cut it back to within two feet and a half of the ground in the spring. Below this cut a certain number of shoots will be produced, from which three will be selected to form the main branches or frame-work of the head. All the others are rubbed off when two or three inches long or sooner. At the end of the season we have a tree with three branches.

Fig 112.

Form of a low standard peach tree, with a stem two feet high and a round, open head.

The *second year* these three branches are cut back full one half their length, and from each we take a shoot to continue the branch, and one to form a secondary branch. The other shoots produced below these are pinched or checked to prevent them from interfering with the growth of the leading branches. In the fall of the year we have a tree with six leading branches, and some bearing shoots below on the older wood.

The *third year* each of these six branches is shortened one half, in order to obtain more secondary branches, and some fruit branches on the lower parts. All young shoots on the old wood, whether fruit branches or not, hculd be cut back one half, or as far as may be necessary, to cause the wood buds at their base to push, and make shoots to bear next year.

The formation of the head goes on as described for two or three years more, when it is complete; for peach trees, properly pruned, do not assume such wide-spreading forms as they do naturally.

The main branches and secondary branches should be at equal distances throughout, and far enough apart to give the bearing wood on their sides the full benefit of the sun and air.

An equality of vigor should also be preserved amongst them by summer pinching. It is not uncommon to see a very vigorous shoot start up in a peach tree, and appropriate so much of the sap as to injure a whole branch; these should be checked the moment their character is observed, unless they may be wanted to fill a vacancy. Every part of the branches should be furnished with bearing shoots, and these should, every spring, be shortened in one half or more, to produce others at their base, whilst those that have borne are cut out.

Some people imagine that when they have taken a pair of hedge shears, or some such instrument, and shorn off the ends of the shoots on the outside of the tree indiscriminately, they are "shortening in," and so they are, as they would a hedge! Some of the shoots are cut away entirely, fruit buds and all, whilst others remain untouched, and the tree becomes like a brush on the outside and naked within. This is almost as bad as the let-alone system. Every shoot should be cut separately. The most expeditious instrument for doing this, is a pair of light hand-pruning shears, such as the French *secateur* (see instruments). A person accustomed to its use can prune every shoot on a full-grown tree in an almost incredibly short space of time, as compared with that required with the knife. Extensive orchardists may be deterred from such a labor, looking to the cost; but if they will engage quick, active, intelligent persons to do the work, and estimate the increased value of the fruit, and longevity and beauty of their trees, there can be no doubt but it will be found a *paying* investment.

Root Pruning.—In gardens where the soil is rich, and trees very full of vigor, disposed to grow too much and bear too little, root pruning should be practised once in two or three years—the first lightly, removing only the ends of the large feeding roots. The safest time to do it is between the fall of the leaf and the opening of spring. Vegetation in the peach seldom becomes sufficiently inactive during the growing season, to enable the roots to be pruned with safety.

The Peach in the form of a vase.—Among al. the forms in which trees are conducted, this is, when well done, one of the most graceful.

It consists of a short stem two to five feet, according to fancy, with a head composed of three or four main branches, and two or three times that number of secondary branches, all trained by means of light stakes at first, and afterwards wire or wooden hoops in the form of a vase or goblet. The branches are arranged in a circle, with bearing shoots filling up the spaces. No shoots are permitted either in the interior or in front that is projecting from the exterior surface of the goblet.

The most beautiful trees of this form are to be seen in the gardens of the Luxembourg, at Paris, and elsewhere in France.

Mr. Louis Gaudry, who has a very pretty little plantation in Paris, and who has published a small work on pruning and training trees, gives the annexed cut as a representation of one of his vase peach trees of eight years' growth (fig. 113). The following is the substance of his mode of conducting them.

First Pruning.—The stem of the yearling tree is cut back to the point at which it is desired to commence the head to three buds, forming a triangle and as nearly as

234 PRUNING.

possible of the same height. Three shoots are obtained from these three buds to form the first or main branches

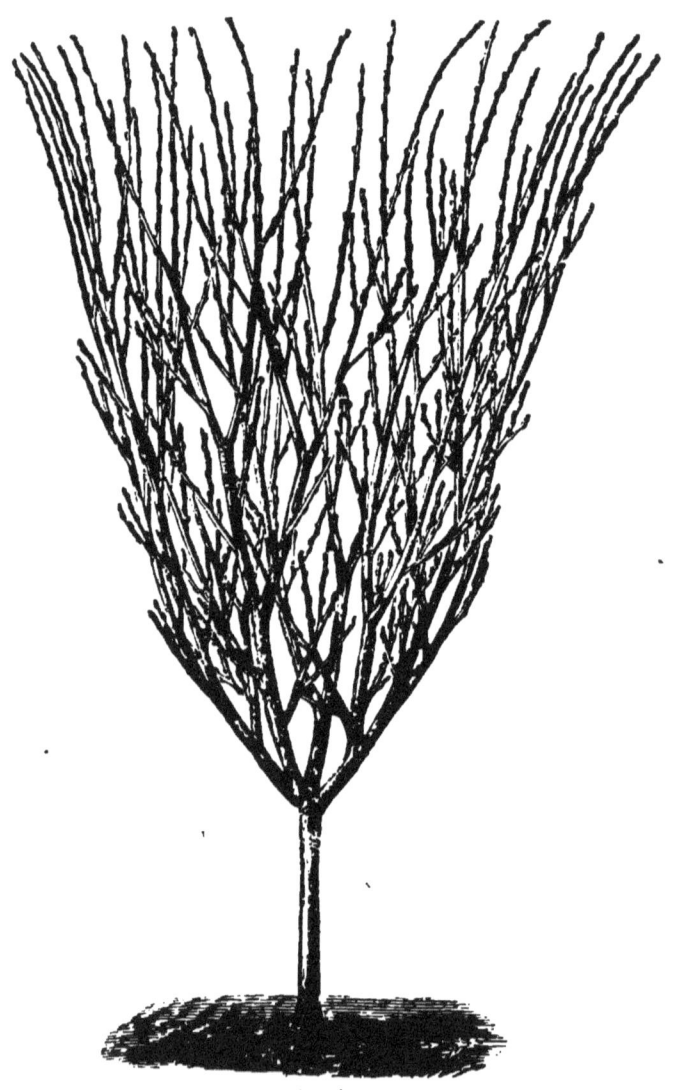

Fig. 113.

Peach tree in the form of a vase, with four main branches, each having several secondary branches. The stem in the figure is five feet, but should not exceed two.

or frame-work of the vase. To favor the growth of these, all the shoots produced below them are rubbed off.

In order to give them the proper inclination, three small stakes are inserted in the ground, to which the three branches are fastened; it is supposed that if these stakes be sunk as far from the base of the tree as the roots extend, and in an upright position, there will be a sufficient opening or space in the centre. The branches should be thus brought out about August, so that the formation of new layers of wood subsequent to that time may fix them in their places. The side shoots, which are produced on the young branches, towards the latter part of the season, designated by the French *bourgeons anticipés*, are pinched to one or two leaves.

2d. *Pruning.*—The spring following, the branches are loosed from the stakes, and shortened to six or eight inches of their base, to a leaf bud on the outside or front of the branch, and with a bud below it, either on the right or left side. The front bud continues the main branch, and the side bud forms a secondary branch. The three branches are pruned in this way, taking care that the secondary branch on each is on the same side, so that two of them cannot come in contact. To favor the growth of these new shoots, all those situated below them that acquire too much vigor, must be pinched at three or four leaves.

A wooden hoop may now be placed in the centre, to which the branches are attached to keep them in their places. In this way the tree progresses; every year one or more secondary branches are produced, the main branches increase in length, and fruit shoots are produced on all the intervals of the branches, on their two sides.

All shoots that push either inside or in front of the vase are pinched off, and pinching is practised at all times to maintain equal growth between the different parts, and to check any too great tendency of the sap to the extremities.

Third pruning.—The fruit branches are pruned to three or four buds, to induce the lower wood buds to push and form new wood for the next season.

The main branches are cut back to ten or twelve inches above the previous pruning, to a bud on the front to continue the branch; the buds selected to produce another series of secondary branches, must all be on the side *opposite* the previous ones. If the position of the buds renders this impossible, then they may all be chosen on the same side as the first.

The hoops this year will require to be larger in diameter than the preceding, in order to give increased width to the vase as it proceeds upwards. All the other operations are conducted in the same manner. The hoops inside are placed within six to eight inches of one another, and the circular branches within twelve to fifteen inches. As the tree advances in age, the growth may become too vigorous at the top; and in this case, the main branches, always the most vigorous, must be pruned short, and even pinched during summer, to turn the sap to the benefit of the weaker parts.

These are the main points in the management of these vases. It may be added, that the apple, pear, cherry, and indeed all other trees may be grown in this form, and by the same means, varying it only to suit different modes of growth and bearing, and degrees of vigor.

The Peach as an Espalier.—Espalier training will never be practised in this country to any very great extent, and therefore it may be considered, in comparison with open ground systems, unimportant. Yet there are some districts not so favorably situated as to be able to produce peaches, apricots, and nectarines, in the open ground. For these a proper system of espalier training is important, because in this form trees are easily protected from winter or spring frosts, and they ripen their fruits perfectly, where open ground or standard trees would not.

The Peach as an Espalier trained on a wall or trellis.—There are a multitude of forms for espalier trees where training on walls or trellises is necessarily and extensively practised, as in England and France. The great requisites in a wall tree are, *first*, to have all the wall covered; and, *second*, to have the different parts of the tree alike favorably placed, with reference to its growth. Next to these are simplicity and naturalness.

The most popular form in England is that called the

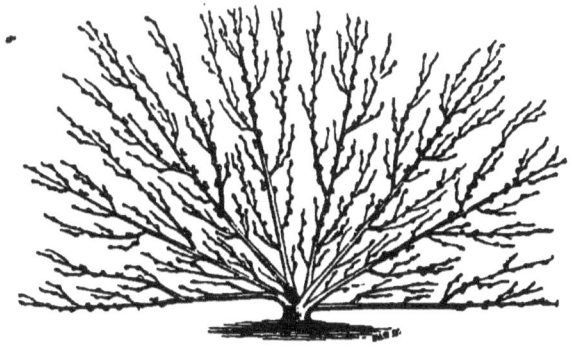

Fig. 114.
Fan-shaped Espalier.

fan (fig. 114). In it the branches are spread out so as to resemble a fan; the lower ones are nearly or quite horizontal; the next more oblique; and so they proceed until the centre ones are quite upright, and this appears to be the defect of this form; for the horizontal branches cannot maintain such a vigor as those more erect above them. The *square* espalier, invented by M. FELIX MALO, of France, and now extensively practised by some of the best peach growers of the celebrated town of Montreuil, seems to possess more advantages, all in all, than any other. The "*Bon Jardinier*," from which the following description of the method of conducting these trees is taken, says: "This generally approved form begins to find imitators, and it is probable that one day it will be adopted by all intelligent gardeners."

First year.—We will begin with a peach tree one year from the bud, and cut it down to within six or eight inches, or three or four buds of the stock. From the buds produced below the cut, two of the strongest are chosen, one on each side, to form the two main branches—*branches mere;* all the other shoots are destroyed, and these two are allowed to grow upright, and in the fall they will be three to four feet high.

Second year (fig. 115).—In the spring, when hard frosts are no longer apprehended, the branches are examined to see if they be sound and healthy, free from bruises, insects, etc., and they are cut back to twelve or fifteen inches of their base, according to their strength; a weak branch ought always to be cut back in such a case as this further than a strong one.

Fig. 115.
Second year.

Fig. 116.
Third year.

The bud cut to, should, if possible, be on the *inside*, and the next bud below it on the *outside;* the first to continue the main branch, and the other to form the first exterior secondary branch. All shoots starting on the front or rear of the main branch should be rubbed off, and those on the sides laid in early to prevent their acquiring too much vigor. The main branches are left till July, when they are brought down to the form of a V, and attached to the wall or trellis in this position. The exterior secondary branch is placed more oblique, and the fruit branches are kept in a uniform and moderate growth by pinching and laying in. The most vigorous should always be laid in first to check them, and favor the others.

Third year (fig. 116).—After loosening the tree from

the trellis, the two main branches are cut back to sixteen or eighteen inches of the previous pruning, and the two lower or secondary branches to twelve or sixteen inches.

The fruit branches are shortened to within two or three buds of their base, and all are again fastened back in their places. When the young shoots have reached the length of three, four, or five inches, such as are badly placed on the front or rear of the branches, or in any place injurious to the symmetry of the tree, are removed. During the summer the different branches must be laid in from time to time, the most vigorous first. This year two more secondary branches must be obtained on each side, in the same manner as in the previous year. Their growth is also promoted by the same means.

The fruit branches on the sides of the main branches may give a few fruit this year, and those on the secondary branches may bear next year.

The fruit branches that have borne are to be cut away each year and replaced by others, therefore we must commence to provide for these branches of replacement. They are produced as follows:

First, it may be observed that fruit branches have generally one or more wood buds at their base. Sometimes these will push and form branches of replacement without any assistance, more than cutting back. In such a case there is no difficulty. When the fruit is ripe, or at the next pruning, the fruit branch that has borne is cut away, and the new one takes its place. But nature does not always act thus. It is generally necessary to force the development of these branches of replacement, withou which the branches in all their lower parts would becom' entirely denuded.

Hence, then, when a branch of replacement fails to appear by the ordinary method of shortening, we have two modes of forcing it: one is to make, after the fruit is set, an incision through the bark two inches above one of the wood buds, and pinch close all the shoots on the fruit branch, leaving only rosettes of leaves necessary to the perfection of the fruit; pinching must be repeated all the time that the shoots on the fruit branch continue to grow.

Fourth Year (fig. 117).—After having examined if the tree is equally vigorous in all its parts, and having decided upon the means of restoring the balance if it has been lost, the tree is detached from the wall or trellis, and pruned, commencing with the fruit branches that have borne. These, it must be remembered, are to be cut back each year to the new branch of replacement produced at its base. The young shoot then becomes the fruit branch, and is pruned within four to fourteen inches, according to their vigor and the situation of the fruit buds.

The two main branches are cut back to within about twenty inches of the previous pruning; the first shoot on the inside is chosen to continue the branch, and the next one below it, on the lower and outer side, to produce the third exterior secondary branch.

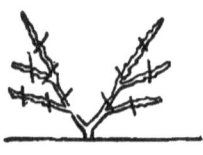
Fig. 117
Fourth year

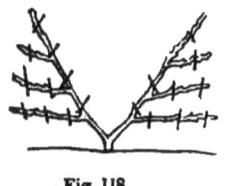
Fig. 118.
Fifth year.

The two secondary branches already formed are cut back to about twelve to fifteen inches of the previous pruning, in order to make all the lateral buds on them push. The terminal bud produces a leader to continue them; all the others are fruit branches.

In attaching the tree again to the wall, the angle that exists between the two main branches is gradually widened, the branches a little more spread at every pruning.

Fifth year (fig. 118).—The tree is now composed of two main branches, both of which have three secondary branches on their exterior lower sides, and fruit branches on all their length on the interior and upper side; and all that is wanted to complete it, is to transform three of the bearing shoots on the upper sides into three secondary branches, corresponding and alternating with the three lower ones. To do this, we select the fruit branch on each, nearest the fork or base of the main branches. The growth of this is favored by training it in an upright position, and by pinching any vigorous shoots near it. The tree is managed thus, as in preceding years, in regard to laying in the shoots according to their vigor, and pinching to maintain regularity, &c.

The sixth year (fig. 119).—The pruning is conducted on the same principles precisely, and another interior secondary branch is produced in the same way as last year.

The seventh year (fig. 120).—Another is produced on

Fig. 119.
Sixth year.

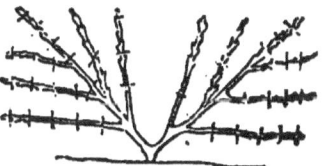

Fig. 120.
Seventh year.

each, and then the tree with its two main branches, and twelve secondary branches, all trained in the form of a parallelogram is complete (fig. 121).

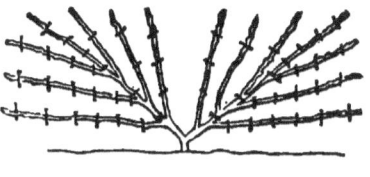

Fig. 121.
Eighth year.

Fig. 122 represents the tree complete, bearing shoots and all.

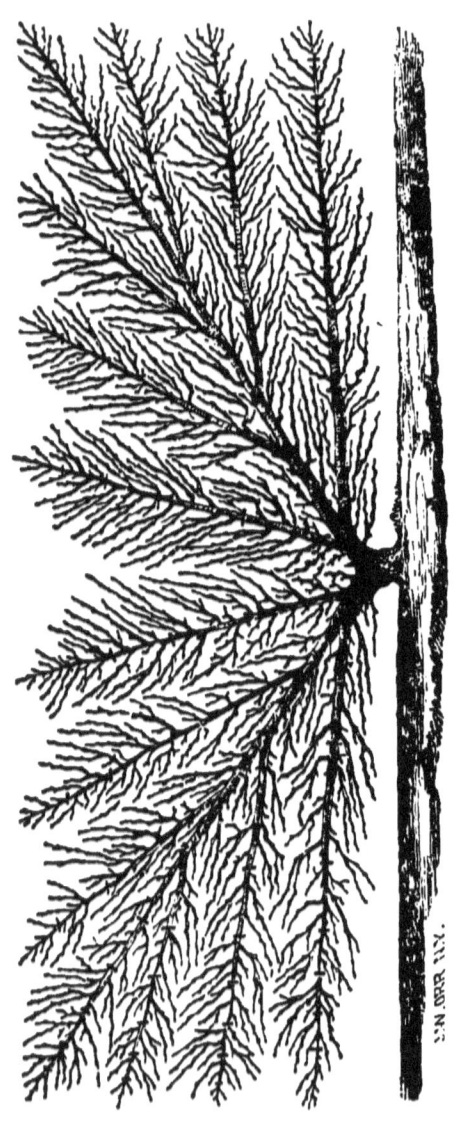

Fig. 122.

The main branches should be permanently fixed at an angle of 45°. The lowest exterior secondary branches at 15°. Some cultivators recommend that the interior secondary branches converge to the centre at an angle of 45°. This gives them an oblique direction, and places them upon a more equal footing with the other parts.

In training such trees, an imaginary circular line is produced on the wall or trellis, and this is divided off into parts, corresponding to the **degrees** of a circle, commencing at the centre above, and numbering both ways to the base: this enables the persons who conduct the tree, to lay in the branches on both sides at an equal angle with precision, which is quite requisite to maintain uniformity of growth and vigor.

It has been considered necessary to treat this subject somewhat minutely, for the purpose of giving to persons wholly unacquainted with training, some knowledge of the principles on which it depends, and the mode of its execution.

The form described above is one of the simplest of all espaliers, except the horizontal, described in treating of the apple and the pear; but whoever can train a tree in this manner well, can do it in all others, for the principles of growth are the same always; and he who understands these, can mould his trees at pleasure, provided he can bestow the necessary labor. The peach may be grown in any or all the espalier forms.

Laying in, and fastening the trees to walls and trellises.—When trees are trained to a wall or fence, the branches are fastened in the desired position by means of shreds of cloth or list, half an inch wide, and from two to three inches long, according to the size of the branch to be laid in. Very small nails are necessary to train on boards, but larger ones on a brick and stone wall. On a trellis, strings of bass matting are used instead of nails and cloth; and in fastening to simple rails, small willows may be used. The principle to be observed, in laying in and fastening the branches and shoots of espalier trees, is that *strong shoots must be laid* in *sooner than weak ones*, and also more inclined from the vertical direction. A great deal may be done towards maintaining uniformity of growth in the different parts of a trained tree, by laying in the branches in a judicious and discriminating manner.

Section 5.—Pruning and Management of the Plum.

The plum bears its fruit on spurs produced on wood two years old and upwards, like the cherry (see fruit branches). On young trees these spurs are several years in the process of formation; but when they commence to bear they endure, if well managed, for many years. They are generally furnished with wood buds on their lower parts; and when they begin to grow feeble, they ought to be renewed by cutting back. The plum is almost universally grown as a standard, and the head may be conducted in the same manner as described for the cherry. The branches should be mainly regulated by summer pinching, to obviate the necessity of knife pruning, that frequently gives rise to the gum. Some varieties of very rapid growth produce shoots three or four feet long in one season; and if not shortened back at the spring pruning, the tree presents long naked branches in a short time.

The chief difficulty in the way of conducting it as a pyramid, is its great vigor; but this can in a great measure be overcome by the use of *dwarfing stocks*, by *pinching* and by *root pruning*.

The latter will be found a most efficient mode of keeping the trees small and fruitful. We have had no experience with the plum as a pyramid; but Mr. Rivers says, that by root pruning annually in October and November, he has succeeded in making handsome pyramidal trees. Standards and dwarf standards may also be root pruned to advantage in small gardens, and where it is desirable to get them into early bearing.

The plum may be trained in any of the espalier forms already described, and in the same manner.

Section 6.—Pruning and Management of the Apricot.

The apricot, like the peach, has fruit and wood buds mixed on the shoots of one year's growth. It has also little fruit branches or spurs like the plum, which are capable of being renewed by shortening.

The mode of pruning must therefore have in view the production of young wood, and maintaining the spurs in a vigorous. and fruitful state. When neglected, it becomes, like the peach, denuded of young bearing wood in the interior, and enfeebled by over-fruitfulness. The shoots should therefore be shortened every season according to their length, as recommended for the peach, to reduce the number of blossom buds, and favor the production of new bearing wood.

It is very liable to the gum, and severe pruning with the knife should be obviated as far as possible by pinching. It may be conducted as a standard, pyramid, dwarf, or espalier, on the same principle as other trees. When trees become enfeebled by neglect or age, they can be renewed by heading down close to the stem. New and vigorous shoots are immediately produced that form a new tree. This heading down should be done very early in the spring, and the wounds be carefully covered with grafting wax.

It is one of the first of our fruit trees to blossom in the spring, and therefore in some localities the flowers are killed by the frost. Where this is apprehended, it may be well to plant on the north side of a wall, or something that will rather retard the period of blooming, and subject it less to freezing and thawing. We have apricots trained here on a south aspect, yet in seven years the blossoms have not been killed, though in one or two instances they have been slightly injured. The espalier trees offer great facility for protection; and therefore, where spring frosts prevail, the apricot should be so trained. Mats or straw hurdles can be placed against them, both in spring and winter if necessary, with the same ease that a common frame is covered.

Section 7.—Pruning the Nectarine.

The nectarine is but a smooth skinned peach. The trees are so similar in their mode of growth, buds, etc., that they cannot be distinguished from one another, and, therefore, whatever has been said respecting the pruning and treatment of one, applies with equal force to the other. This fruit is so infested with the curculio, that it is almost impossible to obtain a crop that will pay for culture in any part of the country in the open ground. Unless some more effective remedy be discovered than any yet known, it will soon have to retire from the garden, and take up its residence with the foreign grape in glass houses.

It produces excellent crops trained in espaliers, on a back wall, or a centre trellis of one of those cold graperies now becoming so popular.

Section 8.—Culture, Pruning, and Training Hardy Grape Vines.

The management of our native grapes is exceedingly simple. Immense crops of Catawba and Isabella, and especially the latter, are raised throughout the country in the entire absence of any systematic mode of training or pruning. A single vine in a neighbor's garden, carried to the flat roof of an outbuilding, and allowed to ramble there at pleasure, without any care but a very imperfect pruning every spring, produces annually many bushels of fruit. But the quality is, of course, greatly inferior to that produced on well-pruned, trained, and dressed vines. A grape vine neatly trained on a trellis, with its luxuriant ample foliage, and rich pendulous clusters of fruit, is really one of the most interesting objects in a fruit garden, and, at the same time, one of the most profitable; for the shade and ornament alone that it produces, are a sufficient recompense for its culture.

In planting a grape vine the first point is to prepare a border for the roots.

This must, in the first place, be perfectly dry. If the soil or situation be wet or damp, it must be drained thoroughly, so that no stagnant moisture can exist in it. In the next place it must be deep—three feet is a good depth; and it must not be less than two where abundant and fine crops are expected. The mode of preparation is, to dig out the natural soil to the required depth, and the length and width necessary. For a single vine, the border should be eight or ten feet long and four wide.

When the excavation is made, if the soil be stiff or damp, a few inches, or a foot deep, of small stones, brick, rubbish, etc., may be laid on the bottom as a sort of drainage. On the top of this deposit the compost for the border. This may consist of two parts of good, fresh, friable loam, one of old, well-rotted manure, and one of ashes, shells, broken bones, etc., all completely mixed with one another. The top of the border, when finished, should be at least a foot higher than the surface of the ground, so that it may still remain higher after settling. Having the border thus prepared, the next point is the *trellis*. The form of this will depend on the situation it is to occupy, and the mode of training to be adopted. Fig. 123 represents one intended for a wall. The principal bars or frame-work are inch and a half boards, three inches wide, nailed together at the angles.

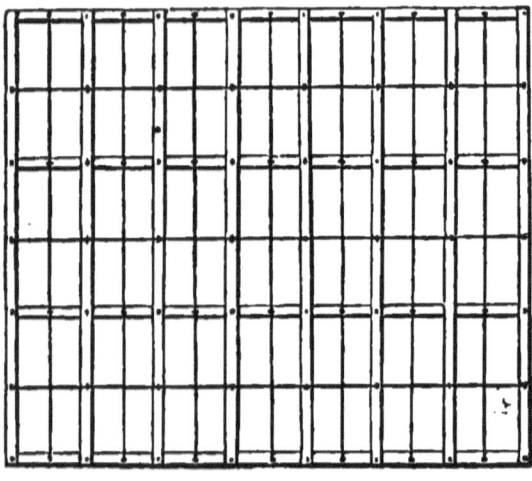

Fig. 123.

Trellis for a grape vine.

It is intended for one vine, and may be the height of the wall that it is intended to occupy. The vertical or upright bars are three feet apart and the cross ones six feet; between them are rods of stout wire. The first or lowest cross bar may be two feet from the ground. It is fastened to the wall by iron hooks or brackets. The best and simplest mode of training a vine on such a trellis as this, is to produce two main branches or arms to be trained in a horizontal manner on the first cross bar. From these two arms, permanent, upright canes are trained, one to each of the upright bars of the trellis. These upright canes produce on their sides a succession of bearing shoots from year to year, being pruned after what is called the "spur" system.

Planting the Vine.—As in planting any other tree, the roots should be carefully spread out, and the fine earth worked well in amongst them. Its position should be exactly in the centre of the trellis it is to be trained on.

Pruning.—It must first be observed that the grape vine bears its fruit on shoots of the current year, produced from eyes on the previous year's wood. Fig. 124 represent the old wood, with its bearing shoot. It is im-

Fig. 124.

Fruit branch of the grape. The cross line towards the points shows where it ought to be stopped.

portant to understand this, because it shows the necessity of keeping up a supply of young wood wherever we desire fruit to be produced.

To illustrate the pruning, we will suppose the plant to be one or two years old, as ordinarily sent out from the nursery. It may have only one shoot, or it may have several. However this may be, all are pruned off but the strongest, and it is cut back to within two eyes of its base. These two eyes will produce shoots, and when they have made a growth of two or three inches, the weaker one is rubbed off and the strong one trained up. It is allowed to grow on till September, when the bud is pinched to mature and strengthen it. Any side shoots that appear during the summer, should be pinched off, as well as any suckers that may appear about the roots.

Second Year.—If the shoot of last year made a strong growth of ten or twelve feet, it may be now cut back to three eyes, and two canes be trained up; but if it made only a weak growth, it should again be cut back to two

11*

eyes, and one shoot only trained up. Side shoots and suckers are pinched off during the summer; and in September these canes are stopped as before, and no fruit is allowed.

Third Year.—We have now two strong canes with which we commence the frame-work of the vine. Each of these is cut back at the winter pruning to within two or three feet of its base, and laid in, as in fig. 125, and fastened to the lower horizontal bar of the trellis. The

Fig. 125.

Grape vine at the beginning of the second year. The arms shortened at *c*, *a*, *b*, etc., are buds.

bud on the end of each at *c*, will produce a shoot to continue the prolongment in a horizontal direction, and a bud (*a*) on the upper side of each will produce a shoot to be trained to one of the upright bars—the first one on its division, or half of its trellis; all others are rubbed off, or the buds cut out. Thus each of these arms produces two shoots—an upright and a horizontal one. During the summer, these shoots are carefully tied in as required, and side shoots and suckers pinched off when they appear. They are also topped in September, as before.

Fourth year.—Each of last year's shoots is cut back to within three feet of its base. It may be necessary to cut the horizontal ones closer than the upright ones, to obtain another strong upright shoot. The two upright canes already established, will produce a shoot from their tops, to continue their extension upwards, and the horizontal ones, as before, produce a shoot at the point to be carried outwards, and one on the top to be trained up to one of the upright bars. This year, several fruit shoots will be produced, on each of which, one or two bunches of grapes may be ripened. In this way the vine goes on adding every season two new upright canes, and two or three feet in length to the previous ones, until the whole trellis is covered; when the management will consist in pruning the spurs every winter to about three eyes. Each fruit branch should only be allowed to produce two bunches of fruit, and the top should be pinched at the second eye, or joint above the fruit (see cross line, fig. 124), in order to arrest the production of useless wood, and turn the sap to the benefit of the fruit. Fig. 126 represents the appearance of a vine trained in this way.

By such a system as this the trellis is covered in every part with bearing wood, the fruit and the foliage are all exposed fully to the sun, an uniformity of vigor is maintained between the different parts, and the appearance is beautiful.

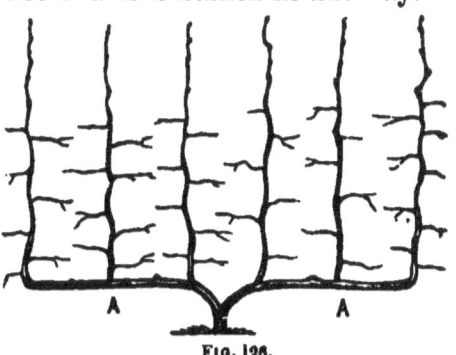

FIG. 126.

Trained, with horizontal arms, *A*, *B*, supporting vertical permanent canes, spur pruned.

A trellis may be covered with a vine by other modes requiring less labor perhaps, and less time, but none will be found more beneficial or satisfactory in the end.

In the management of a grape vine, as in the management of other trees, summer pruning is of great consequence. If a vine is left to itself all summer, or from one winter pruning to another, it will be found that a vast quantity of useless wood has been produced, and that to the serious detriment of the bearing shoots for the following year. Every two weeks the growing vine should be visited, shoots tied in, strong ones checked, superfluous ones rubbed off, and every part kept in its proper place, and in a proper degree of vigor. In certain cases, where the mode of training above described cannot be conveniently adopted, two or three poles, twelve to fifteen feet high, may be sunk in the ground, with a space of three or four feet between them at the bottom, and fastened together at the top, forming a cone, around which the permanent canes may be trained in a spiral manner.

This produces a very beautiful effect, and occupies comparatively little space, but the grapes will not all ripen so well, nor will the training be so easy as on the flat surface of a trellis.

Very tasteful arbors may also be made over some of the walks, by training the vine over the woodwork, in the same manner as on a trellis.

This is a very common practice and offers many advantages. Ingenious persons who care well for their garden, as well in its appearance as its productions, will conceive other plans still better adapted to their particular wants and taste than any of these; but the main point must always be kept in view, that is, to provide for the foliage and the fruit, a free open exposure to the sun. Any system that does not secure this, will fail to a greater or less extent.

The Isabella grape succeeds well even as far north as Maine, by laying it down in winter and covering it with mats, straw, boughs of evergreens, &c.

Vineyard culture.—Vineyards are located on dry sunny hill sides; the land is deeply trenched with the spade or subsoil plough (generally the former, as it is more thorough), and liberally manured. The vines are planted in rows, six to eight feet apart, and four to six feet apart in the rows, and are trained to oak or cedar posts, six to eight feet high. The young vines are cut back close for the first year or two, until they have become well rooted and strong, and only one shoot is allowed to grow. About the third year, one shoot, six feet long or so, is left to fruit, and a new shoot is carried up that season to bear the next. At the following pruning the cane that bore is cut away, and thus a continual succession is kept up. During the summer suckers and superfluous shoots are kept down, and the ground is kept in good clean condition with a horse cultivator principally. As the vines grow old, two and sometimes three bearing canes are taken from each stool.

The vineyards of Cincinnati cover several hundred acres, and from the Catawba grape they make a " sparkling champagne," as good as the French. This is destined to be an important branch of culture.

Culture of foreign Grapes in cold vineries.—Repeated experiments made during many years in all parts of the country, have convinced people generally that the delicious varieties of the foreign grape cannot be produced with any considerable degree of success in the open air. A large number of the hardiest French and German sorts have been tested in our ground, but not one of them has borne satisfactorily. A few good bunches have been obtained the first season or two under very favorable circumstances; but after that the failure is complete. This has rendered glass, heat, and shelter necessary.

The building.—These are constructed of all sizes and at various degrees of expense, from $50 to $500. Some have single lean-to roofs; others have double or span roofs. The walls of some are built of brick or stone; others are of wood, wholly. The cheapest and simplest structure of this kind is the lean-to. The back may be nine or ten feet high, composed of strong cedar posts six feet apart, and boarded up on both sides. The ends are made in the same manner. The front may be two feet high, or three, made of posts, and boards or planks, same as the back. Sills or plates are put on the front and back walls, and then rafters at three and a half to four feet apart. The sashes slip in between the rafters, and rest on a strip of wood on their sides. Unless the grapery be very small, the sash should be in two parts, the lower one twice as long as the upper, and fixed. The upper to slide down over the under one on pulleys, to ventilate the house: doors are in each end at the back, and means are provided for admitting air in front by the opening of boards like shutters.

The border is made for the vines outside the front wall, or part outside and part in, twelve to sixteen feet wide, also two or three deep. This is done by digging a trench or pit the length and width; draining it thoroughly, that not a drop of water can lodge about it. Then lay a few inches of small stones, broken bricks, shells, etc., in the bottom for drainage; and fill up the remainder six inches above the level of the ground, and sloping outwards, with a good compost, of one half surface loam (turf from an old pasture), and the other of well rotted stable manure, shells, street scrapings, a small portion of night soil, offal, etc. All these must be prepared by frequent turning and mixing a few months beforehand.

The vines may be one or two years old, and are preferable in pots raised from single eyes. They should be planted in the spring. A plant is placed under each rafter outside, and carried through under the wall into the house. The stem is cut back to two or three eyes, and when these break the strongest shoot is selected, and the others pinched off. This shoot is trained, as it grows, to a light trellis of iron, or thick wire rods attached to the rafter, and eight or ten inches from the glass. If all goes well, it reaches the top of the house that season. In September the top may be pinched to check the flow of sap to the point, and throw it more into the lateral buds to increase their strength. During the summer no other shoot is allowed to grow but this.

Pruning.—In November or December it is taken down, pruned, if according to the spur system, which is the simplest, to within three or four feet of its base, laid on the ground, and covered with leaves, evergreen boughs, or mats. There it remains till the buds begin to swell in the spring, when it is again fastened to the trellis. The shoot from the terminal bud continues the cane, and no fruit is allowed on it. Those below it produce lateral shoots, from each of which a bunch of grapes may be taken; and each of these must be stopped at two eyes above the bunch; and this is repeated as often as necessary, to give the fruit the whole benefit of the sap. The leading shoot is again stopped in September by pinching off its point, to increase the vigor of its lateral buds. In the fall, when the leaves have dropped, the vine is again taken down. The leader is pruned back to within three to four feet of the old wood. The laterals that have borne are pruned to three eyes, and it is then covered up. This is the routine of spur training. In *long cane* pruning, the young shoot, after the first season's growth, is cut back to three eyes, and the next season two shoots are trained up. The next season the strongest is selected for fruit, and pruned to about three feet; each of the eyes left will produce a fruit shoot, from which one bunch only will be taken. The weaker cane is cut back to one eye, and this produces a shoot for next year's bearing, and so this goes on. When the vine becomes strong, several bearing canes may be provided for every season. This renewal or long cane is very simple, and requires much less cutting than the spur. It also produces a superior quality of fruit, but in general not so large a quantity.

Thinning the Fruit.—When the fruit attains the size

of a garden pea, one third of the smaller ones should be cut out carefully with pointed scissors (see implements) that are prepared for this purpose. The object of this is, to allow the fruits to swell out to their full size. Varieties that produce very compact bunches require more severe thinning than those of a loose, open bunch.

Cleaning the Vine.—At the time the vines are taken from their winter quarters and trellised, they should be well washed with a solution of soft soap and tobacco water, to kill all eggs of insects, and remove all loose bark and filth that may have accumulated on them during the season previous. The house, too, should be cleaned and renovated at the same time.

Syringing the Vines and the Fruit.—Every one who has a grapery must be provided with a good hand syringe, for this is necessary during the whole season. As soon as they begin to grow, they should be occasionally syringed in the morning, except while they are in bloom. After the fruit has set, they should be syringed every evening, and the house kept closed till the next forenoon when the sun is out warm.

Regulating the temperature.—When the temperature exceeds ninety to one hundred degrees, air should be admitted at the top, and, if necessary, at the bottom.

To prevent mildew.—This may be looked for in July. Syringing freely night and morning, and the admission of air during the warmest hours of the day, are the best preventives of this disease. Mr. Allen recommends dusting sulphur on the floor, at the rate of one pound for every twenty square feet; and if it continues to increase, to syringe the vines in the evening, and dust the foliage with it.

Mr. Buist recommends a solution of five pounds of flour of sulphur in four gallons of water, and after it has settled to add one fourth of it to the water used in syringing.

This is but an imperfect outline of the management of a cold grapery. Those who wish full information on all points of the subject, should consult Allen's excellent work, which treats of all kinds of graperies and their management in complete detail.

SECTION 9.—PRUNING AND TRAINING THE FILBERT.

The filbert in this country is a neglected fruit. It is seldom found in the garden, and more rarely still in a prolific, well-grown condition. Of all other trees, it requires regular and proper pruning to maintain its fruitfulness. The blossoms are monœcious—that is, the male organs which are in long catkins (fig. 36), are produced from one bud, and the female flowers from another.

The blossom or fruit buds are produced on shoots of one year's growth, and bear fruit the next. The fruit is borne in a cluster on the end of a small twig produced from the bud bearing the female organs.

It is said that in the neighborhood of Maidstone, county of Kent, England, the filbert orchards occupy several hundred acres, and from these the principal supply of the London market is obtained. One acre has been known to produce £50 sterling, or $250 worth, in one season. The pruning of these Kent growers is supposed to be most perfect of its kind, especially for their soil and climate. It is described as follows in the "Transactions of the London Horticultural Society:"

"The suckers are taken from the parent plant generally in the autumn, and planted in nursery beds (being first shortened to ten or twelve inches), where they remain three or four years. They are slightly pruned every year, in order to form strong lateral shoots, the number of which varies from four to six. But though it is the usual practice to plant the suckers in nursery beds, I would advise every one to plant them where they are to remain, whether they are intended for a garden or a larger plantation; and after being suffered to grow without restraint for three or four years, to cut them down within a few inches of the ground. From the remaining part, if the trees are well rooted in the soil, five or six strong shoots will be produced. Whichever method is practised, the subsequent treatment of the trees will be exactly the same.

"In the second year after cutting down, these shoots are shortened; generally one-third is taken off. If very weak, I would advise that the trees be quite cut down a second time, as in the previous spring; but it would be much better not to cut them down till the trees give evident tokens of their being able to produce shoots of sufficient strength. When they are thus shortened, that they may appear regular, let a small hoop be placed within the branches, to which the shoots are to be fastened at equal distances. By this practice two considerable advantages will be gained—the trees will grow more regular, and the middle will be kept hollow, so as to admit the influence of the sun and air.

"In the third year a shoot will spring from each bud; these must be suffered to grow till the following autumn, or fourth year, when they are to be cut off nearly close to the original stem, and the leading shoot of the last year shortened two-thirds.

"In the fifth year several small shoots will arise from the bases of the side branches which were cut off the preceding year; these are produced from small buds, and would not have been emitted had not the branch on which they are situated been shortened, the whole nourishment being carried to the upper part of the branch. It is from these shoots that fruit is to be expected. These productive shoots will in a few years become very numerous, and many of them must be taken off, particularly the strongest, in order to encourage the production of the smaller ones; for those of the former year become so exhausted that they generally decay; but whether decayed or not, they are always cut out by the pruner, and a fresh supply must therefore be provided to produce the fruit in the succeeding year. The leading shoot is every year to be shortened two-thirds, or more should the tree be weak, and the whole height of the branches must not exceed six feet.

'The method of pruning above detailed might, in a few words, be called a method of spurring, by which bearing shoots are produced, which otherwise would have had no existence. Old trees are easily induced to bear in this manner, by selecting a sufficient number of the main branches, and then cutting the side shoots off nearly close, excepting any should be so situated as not to interfere with the others, and there should be no main branch directed to that particular part. It will, however, be two or three years before the full effect will be produced. By the above method of pruning, thirty hundred per acre have been grown in particular grounds and in particular years, yet twenty hundred is considered a large crop, and rather more than half that quantity may be called a more usual one; and even then the crop totally fails three years out of five; so that the annual average quantity cannot be reckoned at more than five hundred per acre.

"When I reflected upon the reason of failure happening so often as three years out of five, it occurred to me that possibly it might arise from the excessive productiveness of the other two. In order to ensure fruit every year, I have usually left a large proportion of those shoots which, from their strength, I suspected would not be so productive of blossom-buds as the shorter ones; leaving them more in a state of nature than is usually done, not pruning them so closely as to weaken the trees by excessive bearing, nor leaving them so entirely to their natural growth, as to cause their annual productiveness to be destroyed by a superfluity of wood. These shoots, in the spring of the year, I have usually shortened to a blossom-bud."

Such is the management of these celebrated filbert growers, their principal object being to keep the trees small, open in the centre, and covered in every part with fruit spurs. A similar system, but less severe in the cutting back, may be pursued here; some such course of treatment as recommended for the head of the quince as to form and fruitfulness.

Instead of relying on the spring pruning to subdue vigor and induce fruitfulness, pinching should be practised during the summer; for this not only checks the production of wood, but of roots. Root pruning, too, may be safely practised in August, when pruning and pinching of the branches prove insufficient.

In all cases, suckers must be completely eradicated every season, or as soon as they make their appearance. The want of pruning, and the growth of suckers, make the filbert in nearly all our gardens completely barren; a rank production of wood only is obtained year after year.

We find that grafting the finer kinds on stocks of the common filbert raised from seed, renders the trees much more prolific naturally, and also smaller in size. We have trees here now bearing only three years from the graft; the stems are eighteen inches to two feet high, and they are very pretty. Their natural vigor is greatly subdued by the graft. The French conduct them in pyramids with great success, on the same principle as other trees.

SECTION 10.—CULTURE, PRUNING, AND TRAINING OF THE FIG.

In the Northern States the fig is cultivated with very little success in the open ground, but fine crops are produced in the vineries recommended for foreign grapes; and it is in these only that its culture can yield any considerable degree of satisfaction, north of Maryland at least.

Propagation.—The surest and best mode is by layers. A large branch may be layered in the spring, and will be sufficiently rooted in the fall to be planted out. Cuttings also strike freely, and make good plants in one season. All the modes of propagation recommended for the quince, may be applied to the fig. Cuttings are generally preferred in the South.

Soil.—It succeeds in any good rich, warm garden soil, suitable for other fruit trees. In very light or dry soils the fruits fall before maturity, as they require at that season in particular a large amount of moisture; but it is better that it be too dry than too moist, for in the latter case nothing but soft unripe and unfruitful shoots is obtained, whilst in the former moisture can be supplied at the time when it may be required. The wood should be short-jointed, the buds not more than one-fourth of an inch apart. In England dry chalky soils produce the finest crops.

Pruning.—The fig is somewhat peculiar in its mode of bearing. No blossoms appear, but the figs are produced on the stem, appearing at first like buds. The young shoots of last season bear fruit the next; and the shoots produced during first growth produce fruit the same season, and this is called the "second crop." These never ripen, and should never be encouraged where the plants require protection. In warm climates, as in some of our Southern States, these two crops ripen perfectly, though the first from the previous season's wood is larger and better.

This mode of bearing shows that little pruning is necessary, beyond the cutting away of old or worn out branches, and thinning and regulating others. Unfruitful trees, in a moist and rich ground, should be pinched in summer to check their growth, and concentrate the sap more in the lateral buds. *Root Pruning*, too, may be applied as on other trees. Mr. Downing recommends this in his Fruit and Fruit Trees.

Training.—Wherever the trees are hardy enough to withstand the winter without protection, they may be grown in the form of low standards, as recommended for the peach; but when protection is required, where the branches have to be laid down and covered during winter, they must be grown in stools or bushes, with a dozen or more stems rising from the socket. These are easily laid down and covered, and easily brought up to their places again, in the way that raspberry canes are managed. To produce this form, the young tree is planted in the bottom of a trench about a third deeper than in ordinary cases, and a basin is left around it. At the end of the first season's growth, it is cut back to a few inches of the base; there a number of shoots are produced. As these grow up the earth is drawn in around them, to favor the production of other shoots at their base; and in this way it is managed until the requisite number of branches is obtained.

Protection.—Trained in this way, a trench is opened for each branch, or three or four may be put in one trench, if convenient; they are fastened down with hooked pegs as in layering, and covered with a foot of earth, which should be drawn up in the mound form, to throw off the water.

Ripening the fruit.—In fig growing countries, and to some extent here, there is a practice of applying a drop of olive oil to the eye of the fruit, to hasten its maturity. This is usually done by means of a straw.

Training in Graperies.—The back wall of a lean-to cold vinery is an excellent place for the fig. It may be trained on a trellis in the fan or horizontal manner, but severe pruning must not be practised to produce regularity.

SECTION 11.—PRUNING THE GOOSEBERRY.

The gooseberry produces fruit buds and spurs on wood two years old and upwards. Fig. 127 represents the two-year-old wood, *A*, with fruit buds *C*, *C*, and *B*, the one-year old wood with wood buds, *D, D*. Of these wood buds, the upper one next season would produce a shoot, and the lower ones would probably be transformed into fruit buds. At the base of one of the fruit buds, *C*, may be seen a small wood bud *d;* this during next season will produce a small shoot or spur. The great point to aim at in this country, must always be to maintain a vigorous condition; the moment the plant becomes feeble or stinted, the fruit is so attacked with mildew or rust as to be utterly worthless. Hence it is that young plants usually bear excellent crops for the first or second year, while after that the mildew is in some varieties and situations unconquerable.

The bush should have a stem of three or four inches in height, and a head composed of five or six main branches, placed at equal distances and inclined outwards, to prevent denseness and confusion in the centre. These main branches should be furnished with bearing wood in all their length. The production of such a bush may be accomplished by the following means:

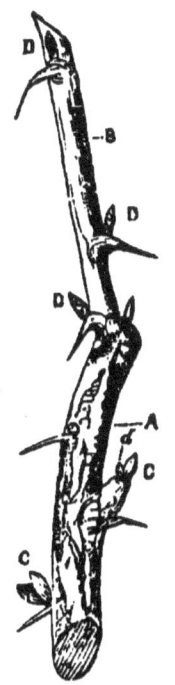

Fig. 127 Branch of the gooseberry, *A*, two year old wood. *B*. one year, *C, C.* fruit buds, *D. D* wood buds, *d*, a small wood bud at the base of fruit bud *C*.

Supposing the young plant as it comes from the nursery to be either a two-year old cutting, or a one-year bedded layer, in either case it will have a stem of two or three inches at least, and a few branches at the top. Before planting, all the buds on the part of the stem to be below the ground are cut out, to prevent them from producing suckers. Among the branches, three of those most favorably situated, are selected for the formation of the head, and the others are cut out entirely. The reserved branches are then cut back to two or three buds; from these one shoot is taken on each branch, and the others are pinched to favor this. By this method we shall have three stout shoots in the fall. If the plant had been well rooted, instead of being newly transplanted, we might have taken two shoots instead of one from each shortened branch. These three branches are cut back at the next pruning to three or four buds, and from each two new shoots are taken, giving at the end of that season six stout young shoots, situated at equal distances. At the next or third pruning these branches are cut back about one-half, in order to produce lateral branches and fruit spurs. At the fourth pruning, the leading shoot is shortened one-third to one-half. Any lateral branches not required to fill up spaces, or such as are improperly placed, are cut back to three or four buds, so as to convert them into fruit branches.

In this way the pruning is conducted from year to year. When the plants become feeble from overbearing, the fruit branches may be headed down and replaced by new vigorous shoots. The better way, however, to provide for this difficulty, is to raise young plants from layers or cuttings, to be at once substituted for such as fall a victim to the mildew. A northern aspect, a cool, damp, substantial soil, and abundance of manure, are all necessary, in connection with the pruning described, to produce fine gooseberries.

The famous growers of Lancashire, England, outdo all the world besides in the production of large gooseberries. The Encyclopædia of Gardening says—" To effect this increased size, every stimulant is applied that their ingenuity can suggest; they not only annually manure the soil richly, but also surround the plants with trenches of manure for the extremities of the roots to strike into, and form round the stem of each plant a basin, to be mulched, or manured, or watered, as may become necessary. When a root has extended too far from the stem it is uncovered, and all the strongest leaders are shortened back nearly one-half of their length, and covered with fresh marly loam, well manured. The effect of this pruning is to increase the number of fibres and spongioles, which form rapidly on the shortened roots, and strike out in all directions among the fresh, newly stirred loam, in search of nutriment.

They also practise what they term *suckling* their prize fruit. By preparing a very rich soil, and by watering, and the use of liquid manure, shading and thinning, the large fruit of the prize cultivator is produced. Not content with watering at root, and over the top, the Lancashire connoisseur, when he is growing for exhibition, places a small saucer of water immediately under each gooseberry, only three or four of which he leaves on a tree; this he technically calls suckling. He also pinches off a great part of the young wood, so as to throw all the strength he can into the fruit.

Section 12.—Pruning and Management of the Currant.

The red and white currants bear like the gooseberry on wood not less than two years old, and, therefore, the same system of pruning may be applied to them. The most convenient and easily-managed form in which they can be grown, is that of a bush or small tree, with a stem of three to six inches high, and a head composed of a certain number, say six or eight principal branches, situated at equal distances, and not nearer to one another at the extremities than six or eight inches.

These branches are produced by cutting back the

young shoots found on the nursery plant, as recommended for the gooseberry. They are afterwards annually shortened to produce lateral branches, when wanted, and fruit spurs. Care must be taken not to prune too close, as this causes the buds on the lower parts to make wood instead of fruit spurs: one third, and in many cases one fourth, will be quite sufficient.

The Currant as a Pyramid.—The currant is very easily formed into pretty pyramids. The mode of conducting them will be similar to that recommended for other trees.

A good strong shoot must first be obtained to commence upon; this is cut back, and laterals produced as though it were a yearling cherry tree. Summer pruning and pinching must be duly put in practice, under any form, to keep up an equality of growth among the shoots, and to check misplaced and superfluous ones. This will obviate a great deal of cutting at the winter or spring pruning. Mr. Rivers, in his "Miniature Fruit Garden," says: "A near neighbor of mine, an ingenious gardener, attaches much value, and with reason, to his pyramidal currant trees; for his table is supplied abundantly with their fruit till late in autumn. The leading shoots of his trees are fastened to iron rods; they form nice pyramids about five feet high; and by the clever contrivance of slipping a bag made of coarse muslin over them as soon as the fruit is ripe, fastening it securely at the bottom, wasps, birds, flies, and all the ills that beset ripe currants are excluded."

The Currant as an Espalier.— It is sometimes desirable, both to economize space and to retard the period of ripening, to train currants on a north wall or trellis; and this is very easily done with success. We have seen the north side of a neighbor's garden fence completely covered with currants without any system whatever being pursued in laying in the branches. The plants were about five feet apart, and the branches were fastened to the wall in a sort of fan form. The proper way to treat the currant as an espalier is, to produce two strong branches on a stem six to twelve inches high. These branches are trained out in a horizontal manner like two arms—one on each side; and from the shoots which they will produce, as many as are to be had at the distance of six inches from one another are trained in an upright position, as in the grape vine (fig. 121).

These upright shoots are managed in the same way as the branches of a bush; they are annually shortened back a little to ensure a good supply of fruit buds.

The *black currant* produces its best fruit on the wood of the preceding year, therein differing from the others. In pruning it, the young wood must be preserved, and branches that have borne must be cut back to produce a succession of new bearing wood, as in the filbert.

Manuring.—No other fruit tree is so patient under bad treatment as the currant, and yet none yields a more prompt or abundant reward for kindness. In addition to the annual pruning described, the bushes should receive a dressing of old, well-prepared manure, two or three inches deep, spread all around as far as the roots go, and forked lightly in. It is a great feeder, and, without these annual dressings, the soil becomes so poor that the fruit is really not worth gathering.

PLANTING, PRUNING, AND TRAINING THE RASPBERRY.

Planting.—The raspberry succeeds well in all good garden soils. The most advantageous and economical position for a raspberry bed in the garden, is generally in the wall border, facing north. In this situation the fruit ripens sufficiently, and the canes are not so liable to suffer from alternate freezing and thawing in the winter. The young canes or suckers are shortened full one half, and planted at the distance of two or three feet. Any flowers that make their appearance on them the first season should be removed, in order to turn all the sap to the benefit of the leaves and new roots, and the production of a young cane for the next season.

Pruning.—The stem is biennial—that is, the canes are produced one season and bear fruit the next, and then die. For example, in fig. 128, *A* is the old cane that has borne, and is of no further use. *B* is the young cane produced at its base last season. The fruit buds produce small shoots, *a, a, a,* that bear the fruit. The pruning is very simple; it consists merely in cutting away early in the spring the old cane that has borne. Some people do this as soon as the fruit is gathered, on the ground that the young cane is strengthened by so doing; but this is questionable. It may be, on the whole, safer to leave it to finish its natural course, and cut it away at the spring or winter pruning.

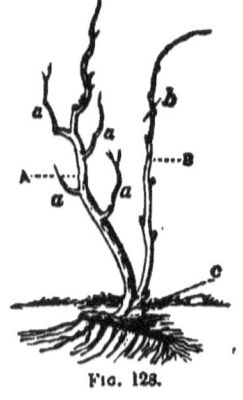

Fig. 128.

The Raspberry. *A*, the old cane that has borne and will be cut away. *B*, the young cane for next season, to be shortened at cross line *b*. *C*, radical bud, to produce a cane next season.

The young cane is shortened to three feet, or three and a half or four, if it be quite stout and vigorous. When the plants have been a year or two in their place, several canes will be produced from one stool in the same season; but three or four only are reserved, and these the strongest. Each one is pruned or shortened as above, in order to concentrate the sap on the bearing buds on the centre and lower parts. This not only increases the size, but improves the quality of the fruit. When the suckers become very numerous, they enfeeble the plant, and it soon becomes worthless. The new ever-bearing variety throws up a great profusion. All the weaker superfluous ones should be carefully removed with a trowel early in the season, say when they have attained five or six inches of growth. In selecting such as are to be reserved, preference should be given to those being nearest in the regular row of plants. Some of the French authors recommend leaving a hole ten or twelve inches deep around each plant at the time of planting, to be filled up gradually, three or four inches a year, with fresh earth, to promote the formation of vigorous radical buds, at the collar of the root, as recommended for the fig.

Manuring.—A liberal dressing of well-decomposed manure should be given them every fall, worked carefully in among the roots with the digging fork. With this treatment a bed will continue productive for seven years at least.

Training.—Mr. Dubriel describes a very pretty and simple method of training practised in France, and I had the pleasure of seeing it carried into practice in the Rouen Garden (fig. 129).

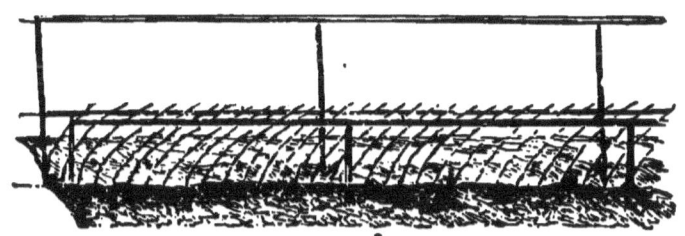

Fig. 129.

French mode of training the raspberry to stakes and ropes.

The railing *B*, is a narrow strip of board, or a small pole, supported on upright stakes; it is eighteen inches from the row of plants, and three feet from the ground. When the young bearing canes are pruned in the spring, they are bent over and fastened to this rail; and thus the young suckers grow up without mixing with the fruit branches; consequently the fruit ripens better and is more easily gathered. During the summer, when the young suckers destined to bear the year following, have reached the height of two feet, they are fastened to a similar rail on the other side of the row, and the same distance from the line of the ground.

The following is an English mode of training described in the "London Gardeners' Chronicle." In fig. 130, the

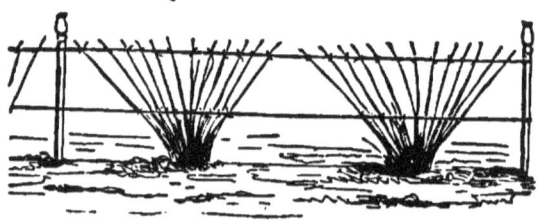

Fig. 130.
English mode of training the raspberry.

uprights between every two or three plants are iron, and the horizontal lines to which the canes are attached, are tarred rope.

In fig. 131, the plants are supposed to be placed in rows four feet apart, and about the same distance from one another in the row. The number of shoots on each

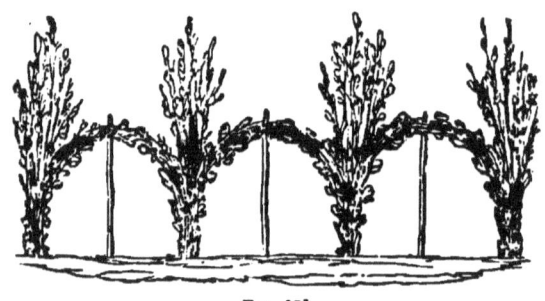

Fig. 131.
English mode of training the raspberry to stakes.

is regulated during the growing season, no more being

allowed to remain than the plant is capable of supporting. In most cases six or eight shoots will be sufficient. Where this method is practised, a row of raspberries in autumn will have something of the appearance represented in fig. 131; the arched portion, tied to the stake in the centre, being the canes which bore fruit last year, and which must be cut down to the bottom, and be replaced by the upright shoots of last summer.

In this last arrangement, five or six fruit-bearing canes are tied together to one stake, and it is impossible that the fruit can either ripen well or be gathered easily. The two first are good and simple plans.

Renovating pyramidal trees of Apples and Pears that have become enfeebled or unproductive by age, bad soil, bearing, or bad pruning.—There are two methods of doing this successfully; one is, to cut back all parts of the tree— the stem may be cut back half its length, the lateral branches at the base to within twelve or fifteen inches of the stem, and shorter as they advance upwards, so that those at the top will be cut to four or six inches. This will preserve the pyramidal form.

It may appear unnecessary to cut back the stem, but we find when this is not done it is almost impossible to secure an equal growth between the upper and lower parts, because the wood at the top is young, and attracts the sap much more than the wood at the base of the old branches below. For a few years after this renewal the young wood at the top must be kept very closely pruned, to prevent it from absorbing more than its due proportion of the sap. When growth commences on trees thus cut back, a large number of shoots will be produced. Amongst those on the stem, a strong and well placed one must be selected for a leader, and its growth favored by checking those around it. Leaders for each of the lateral branches must be selected and encouraged in the same way. The future management will be similar to that described for the formation of young trees. We have succeeded well with a large number of trees thus treated. Where the soil is defective, it must be improved and renewed with fresh soil and composts, so that abundant nutriment shall be given to the new growth.

The second method of renewal referred to is, that of cutting back as already described, and grafting each branch.

The process of regrafting old orchards of standard apple trees, it is well known, renews their vigor, and replaces old worn out and deformed branches with young and vigorous ones, giving to the entire head a healthy and youthful appearance. In many cases this grafting will be much more successful than simply cutting back, for the scions being furnished with young and active buds, that develope leaves at once, attract the sap from the roots, place it in contact with the atmosphere, and carry on the formative process in all parts of the tree with less interruption and greater activity than where reliance is placed upon the production of new shoots on the old wood; for this must be effected by awakening dormant buds, which in many cases takes place slowly, and with more or less difficulty.

SUNDRY OPERATIONS CONNECTED WITH THE CULTURE OF FRUIT TREES.

1st. *The annual cultivation of the soil.*—The soil around fruit trees should, especially in the garden, be kept in a clean friable condition by the frequent use of the hoe and the spade; but in all these operations the roots must not be injured. The forked spade (see implements) is the best for operating about the roots.

2d. *Manuring.*—The very common practice in regard

to the use of manure, is to apply none for several years, until the trees have begun to show signs of feebleness and exhaustion, when large quantities are applied, thus inducing a rank plethoric growth, that can scarcely fail to be seized with diseases. The proper way is to apply a small dressing of well-decomposed material, like some of the composts recommended, every autumn. This should be forked in around the extremities of the roots. There may be rich soils where this will be unnecessary; but most ordinary garden soils require it.

3d. *Mulching.*—This should be a universal practice in our dry and warm summer climate, not only with newly-planted trees, but all, and especially dwarfs in the garden whose roots are near the surface. Three or four inches deep of half decayed stable manure or litter makes a good mulching. It should be applied in May, and remain all summer. After the fall dressing a mulching for the winter will protect the roots and base of the tree from injury; it should be so well decayed as not to attract vermin.

4th. *Watering.*—In dry times, and especially in light, dry soils, fruit trees will derive vast benefits from a liberal syringing over head in the evening, with a hand or garden syringe (see implements). A reservoir in the garden is therefore desirable, and at a point, too, easy of access from all the quarters of the garden. This watering refreshes the trees, drives away insects, mildew, etc., and washes off dust and filth that may accumulate on the foliage and fill up the pores. It is more necessary in city and village, than in country gardens.

Protecting trees against extremes of temperature.- Where the trunk or large branches are liable to injury from sudden changes of temperature in the winter, or from a powerful sun in summer, they may be covered thinly with long rye straw, fastened on with willows.

The trunk alone is more easily protected by means of two boards nailed together, forming an angle for the tree. This is placed on the south side, the injury being induced chiefly by the sun at both seasons.

Newly-transplanted trees, especially if they have tall trunks, and are somewhat injured before planting, may be saved by wrapping them lightly with straw;—a straw rope rolled around answers the purpose. A little damp moss is still better; an occasional watering will keep it cool and moist, and enable the sap to flow under the bark.

PART IV.

SELECT VARIETIES OF FRUITS—GATHERING AND PRE-
SERVING FRUITS—DISEASES—INSECTS—IMPLEMENTS IN
COMMON USE.

CHAPTER I

ABRIDGED DESCRIPTIONS OF SELECT VARIETIES OF FRUITS.

THE accumulation of varieties of fruits within the last ten years has been so great, that anything like a complete description, or account of them all, would in itself exceed the bounds of a moderate sized volume. Taken altogether, there are perhaps at this moment no fewer than *one thousand different varieties* under cultivation. To trace out the history, the peculiar characters and merits of these, must be the work of the pomologist, and forms no part in the design of this treatise. In making the following selections, and in describing them, pomological system and minuteness have not been deemed necessary, nor would they be practicable within the necessary limits.

The main object aimed at is, to bring to the notice of cultivators the *best varieties*, those which ample experience has proved to be *really valuable*, or which upon a partial trial give strong indications of becoming so. Nothing is more embarrassing to the inexperienced cultivator than long lists; and many will no doubt be inclined to think that a large number of the following varieties might very well be dispensed with. But it must be remembered that our country, even the great fruit growing regions of it, possesses different climates, that there are various qualities of soils, various tastes and circumstances of individuals to be provided for. A dozen or twenty sorts of apples or pears may be as many as one person may require; but it does not follow that these varieties *only* are to be cultivated, for it is very probable that another individual, residing even in the same State, would make a selection entirely different. New York, Massachusetts, New Jersey, Pennsylvania, Ohio, Kentucky, and other States, have all *varieties of their own;* and by the time each has made a selection, our one hundred and fifty varieties will be appropriated, and a deficiency still exist in all probability. We are not of those who cry out against new varieties. On the contrary, we look upon every one of *real excellence* as an additional blessing to the fruit growers and to society, for which they should be duly thankful. The only thing to be observed in regard to them is, that before entering into general cultivation they should be fairly and carefully tested under various circumstances. Some well meaning persons make a great cry out against nurserymen and others, whose business it is to experiment, for extending their lists, or noticing new varieties. If such a spirit had prevailed, how would our fruits have been to-day?

It is by no means presumed that the following lists are perfect, even as far as they go. No individual possesses such a thorough knowledge of the various soils and climates of our country, or of the varieties of fruits best adapted to them, as to enable him to recommend with infallible correctness special lists for all localities.

In attempting this, reliance must be placed upon the experience and reports of others, and these are always liable to be biased by tastes or prejudices. These things have been kept in view, and wherever recommendations are made beyond our own knowledge and experience, they are based upon the most reliable authority, and it is hoped will not be found wholly unserviceable to those especially who have neither had experience nor access to sources of extensive and minute information. Those who are not satisfied with the abridged descriptions, are referred to works more strictly pomological; such as "*Downing's Fruit and Fruit Trees,*" "*Thomas's American Fruit Culturist,*" "*Hovey's Fruits of America;*" besides, *Hovey's Monthly Magazine, The Horticulturist, Genesee Farmer,* and other periodicals, where all new and rare fruits are noticed and described.

FIRST DIVISION.—KERNEL FRUIT—APPLES, PEARS AND QUINCES.

SECTION 1.—SELECT APPLES.

CLASS I.—SUMMER APPLES.

1. *American Summer Pearmain.*—Medium size, oblong, skin smooth, red and yellow; tender, juicy and rich. Tree a slow, but erect and handsome grower; bears early and abundantly; one of the best in nearly all parts of the country.—September.

2. *Astrachan, Red.*—Large, roundish, nearly covered with deep crimson, and a thick bloom like a plum; juicy, rich, acid; one of the most beautiful apples. The tree is a vigorous grower with large foliage, and a good bearer.— Russian.—August.

3. *Benoni.*—Medium size, round, red; flesh tender, juicy and rich; a good bearer and strong upright grower. —From Massachusetts.—August.

4. *Bough, Large Sweet* (Large yellow bough of Downing).—Large, pale yellow, sweet, rich flavored. Tree a moderate, compact grower, and abundant bearer.—Aug.

5. *Bevan's Favorite.*—A new Jersey apple, where it is esteemed as one of the best of its season. Medium size, roundish striped, sub-acid and good.—August.

6. *Bohanan.*—A very delicious high-flavored apple

of Kentucky. Roundish, or inclining to oval, deep yellow Very tender, sprightly and fine.—August to October.

7. *Caroline Red June.*—A very early and good apple, cultivated considerably in Michigan, Wisconsin, &c. About as large as the Summer Queen. Have seen good specimens from Kalamazoo.

8. *Early Harvest.*—Medium to large size, round, pale yellow, rich sub-acid. Tree a moderate grower, but erect and handsome, and a good bearer.—Last of July to Aug.

9. *Early Strawberry.*—Medium size, smooth and fair, mostly covered with deep red; tender, almost melting, with a mild, fine flavor. Tree a moderate, erect grower, and a good bearer; a beautiful and excellent variety for both orchard and garden.—Middle to end of August.

10. *Early Joe.*—A beautiful and delicious, small sized, deep red apple. Tree rather a slow, but upright grower, and a most profuse bearer; originated in Ontario County, N. Y.—Last of August.

11. *Early Chandler.*—Medium size, roundish, striped, good quality. Originated in Connecticut, but is extensively cultivated in some parts of Ohio, where it succeeds well. The tree is vigorous and erect.—August and Sept.

12. *Early Pennock.*—This variety is quite popular in some districts of Ohio, and very little known elsewhere. It is described as a magnificent, large, conical, yellow and red apple, ripening there the middle of August.

13. *Garretson's Early.*—Medium size, greenish yellow tender, juicy and pleasant. Tree vigorous, very productive, and bears young. Noticed first in Hovey's Magazine, September, 1848. Supposed to have originated in New Jersey; not much disseminated.—July and August.

14. *Golden Sweeting.*—Large, roundish, pale yellow, a very fair, fine, sweet apple. Tree a strong grower, spreading and irregular; a good bearer.—August.

15. *Keswick Codlin.*—Large, oblong, pale yellow, acid.

Tree erect and very vigorous; bears when quite young and abundantly; excellent for cooking from July to October.

16. *Lyman's Large Summer.*—Large, roundish, pale yellow, rich and excellent. The tree requires shortening like the peach, to keep up a proper supply of young shoots, as they bear only on the ends.—August.

17. *Manomet* (Horseblock Apple).—This is an excellent late summer, sweet apple, originated near Plymouth, Massachusetts, and named by Mr. John Washburn of that town. It is described in Hovey's Magazine, September, 1848, as "one of the finest early sweet apples," "having a deep yellow skin, and a bright vermillion cheek." The tree is vigorous and a good bearer.—August and Sept.

18. *Oslin* (Oslin Pippin, Arbroath Pippin).—A famous Scotch apple, succeeds well in Upper Canada. Medium size, roundish, yellow; flesh juicy, rich and fine. Tree productive.—August and September.

19. *Ornes Early.*—Large, yellow, with a dull, red cheek, tender, juicy and fine. Imported from France to Massachusetts without a name, and afterwards described in Hovey's Magazine under this title. The tree is a strong grower.

20. *Summer Belle-fleur.*—This is a very fine, late summer apple, raised by John R. Comstock, of Duchess County, from a seed of the Esopus Spitzenburgh; it resembles the yellow belle-fleur in form and color, and is described by Mr. Downing, *Horticulturist*, vol. 3, as "decidedly superior to Porter, William's favorite, or any summer apple of its season." Tree strong and upright.

21. *Summer Sweet Paradise.*—A large fine, sweet apple, from Pennsylvania; round, greenish yellow, juicy, sweet and rich.—August and September.

22. *Sine-qua-non.*—Medium size, greenish yellow; flesh tender and fine flavored. Tree a slender, slow grower,

but bears well. Originated on Long Island by the late William Prince.—August.

23. *Summer Rose* (Woolman's Early).—Medium size, roundish, pale yellow, with a red cheek, tender and delicious; has a most beautiful waxen appearance. Tree rather a slow grower, but a good bearer.—Middle to end of August.

24. *Summer Queen.*—Large, conical, striped and clouded with red; rich and fine flavored. Tree grows rather irregular, with a large spreading head.—August.

25. *Sops of Wine.*—Medium size, conical, dark crimson flesh stained with red, tender and delicious. The tree is a fine grower and bearer, and the fruit remains a long time in use; known as the "Pie Apple" about Oswego. —August and September.

26. *Summer Scarlet Pearmain* (Bell's Scarlet) English. —Medium to large, conical, mostly covered with crimson; flesh stained with red; tender and good. Trees grow freely, and bear young and abundantly.—Aug. and Sept.

27. *Williams' Favorite.*—Large, oblong, red, rich and excellent, a moderate grower and good bearer; highly esteemed in Massachusetts, where it originated.—August.

CLASS II.—AUTUMN APPLES.

28. *Alexander* (Emperor Alexander).—A very large and beautiful deep red apple, with a light bloom. Tree spreading, vigorous, and productive. Russian.—October and November.

29. *Autumn Strawberry* (Late Strawberry).—Medium size, streaked light and dark red; tender, crisp, juicy and fine. Tree vigorous, rather spreading, productive; one of the best of its season.—September and October. Cultivated most in Western New York; origin unknown.

30. *Autumn Swaar.*—A large, roundish, flattened, yel-

low apple, generally known as "Sweet Swaar" in the orchards of Western New York. The flesh is yellow, juicy, sweet and rich; tree stout and spreading; very good.

31. *Beauty of Kent.*—A magnificent English apple, rivalling the Alexander in size and beauty, skin striped with dark red; flesh tender but coarse, and indifferent in flavor; excellent for cooking; tree very vigorous and productive.

32. *Bailey Spice.*—A medium-sized, roundish, yellow apple, with a sub-acid, brisk *spicy* flavor, introduced by Jno. W. Bailey, of Plattsburg, N. Y., where it originated; tree of moderate growth, a great bearer,— fruit always fair; little known yet.—September and October.

33. *Cooper.*—A very large, beautiful, and excellent Ohio apple, roundish,—skin yellow streaked with light red; flesh tender, juicy, and agreeable. (Barrels of them were exhibited at the Ohio State Fair in 1850, and nothing in season equalled them. It is said to have been brought originally from New England.)—October to December.

34. *Clyde Beauty.*—Large, conical, pale yellow, striped and marbled with light red; sub-acid, good; introduced to us by Mr. Matthew Mackie, of Clyde, N. Y., where it originated.—October to December.

35. *Duchess of Oldenburg.*—A large beautiful Russian apple, roundish, streaked red and yellow; tender, juicy, and pleasant; tree a vigorous fine grower, and a young and abundant bearer.—September.

36. *Drap d'Or, or Cloth of Gold.*—Large, golden yellow; flavor mild and agreeable; tree spreading.—October.

37. *Dyer.*—See Pomme Royal.

38. *Fall Pippin.*—Very large, roundish oblong, yellow; flesh tender, rich, and delicious; tree vigorous,

spreading, and a fine bearer; esteemed everywhere.—October to December.

39. *Fall Harvey.*—A large, handsome yellow apple, resembling the Fall Pippin, but not so good. Essex county, Mass.—October and November.

40. *Fleiner.*—Medium size, oblong, pale yellow, with a red cheek; tender and pleasant; has a beautiful, smooth, waxy appearance; tree erect, and a great bearer; German.—September and October.

41. *Gravenstein.*—A beautiful, large, striped, roundish apple of the first quality; tree remarkably vigorous and erect in growth, and very productive. German.—September and October.

42. *Garden Royal.*—Small, yellow, striped with red, sub-acid, rich, spicy, and delicious; tree of moderate or slow growth; Massachusetts.—September. First noticed by Manning in Hovey's Magazine. Not adapted for the orchard either in tree or fruit.

43. *Haskell Sweet* (Sassafras Sweet Cole).—Large, flat, greenish yellow, tender, sweet, and rich; tree vigorous and productive. Massachusetts.—September and October.

44. *Hawthornden.*—A beautiful Scotch apple, medium to large size, pale yellow and red; trees have strong shoots, with low spreading heads; constant and abundant bearer; excellent for cooking.—September and October.

45. *Hawley* (Dowse).—A magnificent, large pale yellow apple, mild acid, tender, rich, and fine; tree is a fair grower, and bears well; originated in Columbia county, N. Y.; has black spots in some seasons.—September and October.

46. *Jewett's Fine Red.*—An excellent New England apple, medium size, tender and fine flavored; a good grower and bearer, said to be well adapted to the North.—October and November.

47. *Jersey Sweet.*—Medium size, striped red and green,

tender, juicy, and sweet; a strong fine grower and good bearer; succeeds well, and is highly esteemed in almost all parts of the country, both for table and cooking.—September and October.

48. *Kane* (Cain).—A very beautiful, smooth crimson apple, of Delaware, resembling the Fameuse; medium size, roundish, and good quality.—October.

49. *Lowell, Orange, Tallow Pippin* (Queen Anne in Ohio).—Large, oblong, pale yellow, skin oily, quality excellent; tree a good grower and bearer.—September and October.

50. *Lyman's Pumpkin Sweet* (Pound Sweet).—A large, round, green apple, fine for baking; tree very vigorous, upright, and productive; much grown in Western New York.—October to December.

51. *Maiden's Blush.*—Medium size, flat, pale yellow, with a red cheek, beautiful, tender and pleasant, but not high flavored; tree an erect and fine grower, and good bearer.—September and October.

52. *Munson Sweet.*—Medium to large, roundish flattened, pale yellow, with a blush on the sunny side; tender, sweet, and good; becomes rather dry as it matures; very fine for baking, and very beautiful; introduced to us by Mr. Jesse Storrs, of Marathon, N. Y. It is said to be a native of Massachusetts; tree a good, upright grower, and good bearer.—October to December.

53. *Northern Sweet.*—A very beautiful and excellent sweet apple, introduced to us by Mr. Jonathan Batty, of Keeseville, N. Y., who presented it at the Pomological Convention at New York, in 1849. It is so much like the Munson that we once thought them identical. It is supposed to have originated in Chittenden county, Vt. The tree is a moderate grower, with drooping branches, and a great bearer.

54. *Porter.*—Medium size to large, oblong, yellow

flesh, fine, tender, and of excellent flavor; tree a moderate grower; very popular in Massachusetts.—September.

55. *Pomme Royal, or Dyer.*—Large, roundish, yellowish white, with a brown tinge next the sun, crisp, juicy, and high flavored; tree a fair grower and abundant bearer.—September and October.

56. *Pumpkin Sweet, Pumpkin Russet.*—A very large, round, yellowish russet apple, very sweet and rich; tree a strong, rapid grower, with a spreading head; valuable. —October and November.

57. *President.*—Large and beautiful, yellow, with a red cheek, roundish, flattened, of good quality, bears most abundantly. We obtained it from Columbus, Ohio.—October.

58. *Republican Pippin.*—Large, roundish, striped, subacid; described by Dr. Brinkle, in the Horticulturist, as having a peculiar walnut flavor; " quality No. 1." The tree is a vigorous grower and regular bearer, " but does not bear so well on a limestone soil." September to October, and fit for cooking in July. Originated in Lycoming county, Pa.

59. *St. Lawrence.*—Large, round, streaked red and greenish yellow; a very beautiful, productive, and popular market apple. Originated in Montreal, Canada.— October.

60. *Spice Sweet.*—Large, pale yellow, with a blush on the sunny side, quite waxen and beautiful, tender, sweet, and fine; a great bearer.—September.

61. *Smoke-House.*—This is a fine apple, originated in Lancaster county, Pa. Rather large, flat, striped; tree is a rapid grower, with spreading branches; flesh crisp, juicy, and fine flavored. October. Not much disseminated yet.

62. *Sawyer Sweet.*—Medium size, conical, greenish, with a blush on the exposed side; tender, sweet, and good; the tree is a free upright grower, and a good

bearer. October and November. Described by Kenrick, from whom we obtained it.

63. *Superb Sweet.*—Large, roundish, yellow and red; flesh tender, juicy, rich flavored; tree is a good grower and bearer; native of Massachusetts, and succeeds well in Maine. Described by Cole, who sent it to us.—September and October.

64. *Summer Sweet Paradise.*—Large, roundish, flattened, pale green, marked with gray dots; flesh tender, sweet, and rich; the tree is very productive, and bears young. Originated at Columbia, Pa.; one of the best dessert varieties of its season.—September and October.

65. *Tompkins.*—A large and beautiful apple, from Tompkins county, N. Y., where it is supposed to have originated. It is of a golden yellow color at maturity; flesh sub-acid, tender, and rich; tree productive. October and November. Described in the Horticulturist in 1847.

66. *Towne.*—Large, flat, striped; flesh tender, juicy, mild, and pleasant; ripens with the Gravenstein, and is nearly as good with us; obtained from Kenrick.

CLASS IV.—APPLES FOR ORNAMENT OR PRESERVING.

67. *Red Siberian Crab.*—Small, about an inch in diameter, yellow, with scarlet cheek; beautiful; tree is vigorous and erect; bears when two or three years old.—September and October.

68. *Large Red Siberian Crab.*—Nearly twice as large as the above, but similar in appearance and quality; trees grow large.—September and October.

69. *Yellow Siberian Crab.*—Nearly as large as the last, and of a beautiful golden yellow.

70. *Large Yellow Crab.*—Larger than any of the preceding, pale yellow, with tint of red in the sun; tree a vigorous and rapid grower.

71. *Double Flowering Chinese.*—A beautiful ornamental tree, producing large clusters of semi-double rose colored blossoms.

CLASS III.—WINTER APPLES.

72. *Baldwin* (Steele's Red Winter, in Western New York).—A large, fair, bright, red apple, roundish, inclining to oblong; flesh crisp, juicy, and pleasant; not very fine grained. Tree is a vigorous, rapid grower, with curved erect branches, and forms a regular open head in the orchard; bears abundantly; originated in Massachusetts, where it is one of the most popular and profitable winter fruits for market. It also succeeds well throughout New York, and especially in the Genesee Valley; variable in the south and west.

73. *Baily Sweet* (Patterson Sweet).—A magnificent sweet apple of the largest size, originated in Wyoming county, New York; brought to our notice a few years ago, by E. A. McKay, Esq., of Naples, New York; color deep reddish crimson; flesh tender, sweet, juicy, and rich; tree vigorous, erect, and productive.—October to January.

74. *Bourrassa.*—Large, conical, reddish russet, rich and high-flavored, but rather dry; supposed to have originated in Lower Canada; a very valuable apple for the high northern latitudes; succeeds well in western New York.—October to March.

75. *Blue Pearmain.*—Very large, roundish, purplish red, clouded, covered with bloom; flesh sub-acid, juicy, and good; tree is a vigorous grower, with large foliage, and a spreading head; bears moderately; very popular in the markets on account of its size and beauty.—October to January.

76. *Bell-flower, Yellow.*—Mr. Downing follows Thomp-

son in calling this *Belle-fleur*, which is, no doubt, correct, this being the French term for handsome flower, while our word means bell-shaped flower; but having been described as bell-flower by the older authors, and universally known and called so in this country, it cannot be changed. The fruit is large, oblong, slightly conical, yellow, with a blush on the sunny side; flesh crisp, juicy, pretty acid, and rich; tree is a rapid grower, with spreading and drooping branches, very productive; originated in New Jersey; succeeds well throughout a large portion of the country.—November to April.

77. *Belmont* (Gate).—A beautiful and excellent apple of Ohio; large, roundish, inclining to conical; yellow, with a tinge of red on one side; sub-acid, juicy, and fine; succeeds well in New York and northern Ohio, but is variable at Cincinnati and further south.—October to February. Described by Mr. Downing as "Waxen," he supposing it to be identical with that of Coxe. This is yet in doubt.

78. *Belle et Bonne.*—This is a native of Connecticut; a tree at East Hartford, forty years old, measures one hundred and twenty feet in circumference, and bears forty to fifty bushels a year. Mr. Downing describes it as " a very large, showy, yellow apple, of the fall pippin class;" the tree forms a beautiful symmetrical head, and bears abundantly.—October to January. Keep all winter. There is an apple by this name cultivated about Rochester, large, showy, striped, of fair quality, ripens in September.

79. *Broadwell.*—A fine, new, sweet apple, from Ohio; large, greenish yellow, tender, sweet, and excellent; keeps till spring.

80. *Carthouse* (Gilpin, Red Romanite).—Medium size, round, striped, sub-acid, and agreeable; cultivated rather extensively in some parts of the south, where it is es-

teemed for its productiveness and good keeping qualities It is also grown much for cider.—February to May.

81. *Danvers' Winter Sweet.*—Medium size, roundish, slightly conical; greenish yellow, with a brown tinge on the sunny side; flesh yellowish, crisp, juicy, and rich, tree very productive, spreading; one of the best sweet apples; origin, Danvers, Massachusetts; keeps till April.

82. *Dominie.*—Medium size, flat, greenish yellow, streaked with red, sub-acid, juicy, and high flavored.—November to April. Cultivated rather extensively in the orchards on the Hudson; resembles the Rambo, and like it succeeds well in the west and south.

83. *Dutch Mignonne.*—A very large, beautiful, and excellent apple; a native of Holland; orange, marked with russet and faint streaks of red; fine flavored; tree erect, and good bearer.—November to March.

84. *Fameuse.*—Medium size, deep crimson, flesh snowy white, tender, and delicious; tree vigorous, with dark wood; a beautiful and fine early winter fruit; succeeds particularly well in the north.—November, December, January; September, and October, in Ohio and farther south. Native of Canada.

85. *Fallawater.*—From Columbia, Pennsylvania; medium size, slightly conical, greenish yellow, with a dull blush on the sunny side; flesh juicy, sub-acid, and good; productive.

86. *Fort Miami.*—A new variety from Ohio. Said to be a rich, high-flavored, good keeper.

87. *Green Sweeting.*—Medium size, greenish, tender, sweet, and spicy; one of the very best long-keeping sweet apples; tree a moderate grower.—November to May.

88. *Hubbardson Nonsuch.*—Large, striped yellow and red, tender, juicy, and fine, strong grower and great bearer; native of Massachusetts.—November to January.

89. *Hooker.*—A large fine apple, introduced to Rochester by Judge E. B. Strong, from Connecticut; color greenish yellow, striped, and covered with dark red in the sun; flesh tender, juicy, and fine-flavored; tree very productive.—November to January.

90. *Hartford Sweeting* (Spencer Sweeting).—A native of Hartford, Connecticut; medium size, flat, striped; flesh juicy, tender, and rich; keeps till late in spring; tree very productive; a valuable orchard variety.

91. *Jonathan.*—Medium size, striped red and yellow; flesh tender, juicy, and rich, with much of the Spitzenburg character; shoots light-colored, slender, and spreading; very productive; a native of Kingston, New York. —November to April.

92. *King.*—A large handsome fruit, striped red and yellow, of fair but not first-rate quality; tree remarkably vigorous and fruitful.—October to January.

93. *Lady Apple, Pomme d'Api.*—A beautiful little dessert fruit, flat, pale yellow, with a brilliant red cheek; flesh crisp, juicy, and pleasant; the tree forms a dense, erect head, and bears large crops of fruit in clusters; the fruit sells for the highest price in New York, London, and Paris.—November to May. There are four or five varieties of these described by authors, but this is the best.

94. *Ladies' Sweet.*—Large, roundish, green, and red, nearly quite red in the sun; sweet, sprightly, and perfumed; shoots slender but erect; a good bearer; originated near Newburg, New York; one of the best winter sweet apples.—November to May.

95. *Limber Twig* (James River).—A large, dull, red apple; second rate in quality, but keeps till June or July, on account of which chiefly it is cultivated at the south and west; the tree has weak, pendulous branches, but is exceedingly hardy, and bears immense crops.

96. *Mother.*—Large, red; flesh very tender, rich, and

aromatic; tree a good bearer; succeeds well in the north; supposed to have originated in Worcester county, Massachusetts.—November to January.

97. *Melon* (Norton's).—Large, pale, whitish yellow and vermillion red; flesh tender, juicy, almost melting, and spicy; a most beautiful and delicious fruit; originated in East Bloomfield, New York; the tree is rather a slow grower, but a good bearer; retains its freshness from October to April.

98. *McLellan* (Martin).—Large, roundish, yellow, striped with red; flesh fine grained, mild, tender, and pleasant; tree productive; originated in Woodstock, Connecticut, where it is highly esteemed. Mr. Downing considers it " worthy of a place in every small collection, and valuable for the orchard."

99. *Minister.*—A large, showy, oblong, striped apple, fine-grained and pleasant, sub-acid. Mr. Manning considered it one of the finest apples Massachusetts produced; with us it is about second-rate; trees exceedingly productive.—October to January.

100. *Monmouth Pippin* (Red Cheeked Pippin).—A large, showy, good apple, of New Jersey, somewhat cultivated, and succeeds well in western New York, roundish, oblong, greenish yellow, with a deep red cheek; flesh rather compact, sub-acid, and agreeable; keeps well till March or April; tree upright, vigorous, and productive.

101. *Northern Spy.*—Large, striped, and quite covered on the sunny side with dark crimson, and delicately coated with bloom; flesh juicy, rich, highly aromatic, retaining its freshness of flavor and appearance till July; the tree is a remarkably rapid, fine, erect grower, and a great bearer; like all trees of the same habit, it requires good culture, and an occasional thinning out of the branches, to admit the sun and air fully to the fruit. It

is one of the largest, most beautiful, and excellent long keeping apples yet known; originated in Ontario county, New York, and introduced a few years ago. Mr. Cole says it is found to be very hardy as far north as Maine.

102. *Newtown Pippin.*—One of the most celebrated of American apples, on account of its long-keeping and excellent qualities, and the high price it commands abroad; but its success is confined to certain districts and soils. It attains its greatest perfection on Long Island, and on the Hudson. In western New York and New England, it rarely succeeds well. It requires rich and high culture, and it is said a large supply of lime; tree a slow, feeble grower, with rough bark.—November to June.

103. *Ortley* (Detroit, White Bellflower, Warren Pippin, etc., of the West, Woolman's Long, etc., etc.).—Large, roundish, slightly oblong, pale yellow; flesh sub-acid, sprightly and fine; succeeds well in New Jersey, and in the west; origin, New Jersey.

104. *Peck's Pleasant.*—Large, pale yellow, with a brown cheek, very smooth and fair; flesh firm and rich, approaching the flavor of a Newtown pippin; tree erect and a fine bearer.—November to April.

105. *Pomme Grise.*—Small, greyish russet, very rich, and high-flavored; tree a moderate grower, but a good bearer; very valuable in the north; is frequently shipped from Canada to England.—November to April.

106. *Pickman* (Pickman Pippin).—Medium to large, roundish, slightly flattened, of a beautiful clear straw color; sub-acid and rich; fine for cooking.—November to February. Supposed to be a native of Massachusetts.

107. *Pryor's Red.*—A very popular and excellent fruit in Ohio and Kentucky; somewhat similar to the Bourassa in color and flavor; reddish or brownish russet; rather dry, but rich and high-flavored; rather late and shy

bearer.--December to February. Cultivated extensively for the New Orleans market.

108. *Phillip's Sweeting.*—A new Ohio apple, large, conical, yellow and red; flesh tender, juicy, sweet, and good. Early winter.

109. *Rambo* (Romanite, Seek-No-Farther).—Good and popular over a greater extent of country than any other variety; medium size, round, greenish, yellow, striped with red; flesh exceedingly tender, juicy, and pleasant; tree vigorous, erect, and very productive; ripens in the autumn at the south and west, but keeps here till February.

110. *Rome Beauty.*—A large and very beautiful new apple of Ohio; we saw it at Cincinnati in 1850; it is roundish or very slightly conical, pale yellow, mostly covered with bright red; flesh not very fine, but tender, juicy, and good; early winter. It will undoubtedly be valuable for the orchard.

111. *Rawle's Jannet* (Rawle's Janneting, Never-fail, Rockremain, &c.).—The most popular and valuable orchard fruit of Kentucky. An experienced orchardist of that State, Mr. Sanders, of Carrol county, who has known it for *fifty years*, says that " 30 to 40 per cent. of every orchard in Kentucky should be planted with it." It blossoms two weeks later than most other varieties, and therefore always escapes spring frosts. It is medium to small, round, greenish streaked, and clouded with dull red; flesh compact, crisp, juicy and vinous.—Keeps till June or July.

112. *Reinette, Canada.*—Very large, flattened, ribbed, dull yellow, flesh firm, juicy and rich; tree a strong grower and good bearer. In France it is considered the largest and best apple, and proves excellent here; it keeps better for being picked early.—November to March.

113. *Red Canada* (Old None-such of Massachusetts) —

Medium size, red, with white dots, flesh fine, rich, sub-acid and delicious; tree a slender grower; one of the best of apples.—November to May.

114. *Rhode Island Greening.*—Every where well known and popular; tree spreading and vigorous, always more or less crooked in the nursery; a great and constant bearer in nearly all soils and situations; fruit rather acid, but excellent for dessert or cooking; towards the south it ripens in the fall, but in the north keeps well till March or April.

115. *Russet Golden.*—Medium size, dull russet, with a tinge of red on the exposed side; flesh greenish, crisp, juicy and high flavored; tree a fine grower, spreading with light colored speckled shoots, by which it is easily known; bears well; popular and extensively grown in Western New York.—November to April.

116. *Russet Golden American* (Bullock's Pippin, Sheep's Nose).—Medium size to small, conical, slightly russeted; flesh remarkably tender, juicy and rich; of the finest quality; origin, New Jersey, and succeeds well in the west and south. Tree an upright, compact, but not rapid grower.

117. *Russet English* (Poughkeepsie Russet).—Medium size, slightly conical, greenish yellow, mostly covered with russet. Tree is remarkably stout and erect, bears large crops, and the fruit will keep a year; quality good; very profitable.

118. *Russet, Roxbury or Boston.*—Medium size to large, surface rough, greenish, covered with russet, flavor indifferent; tree vigorous, spreading, and a great bearer; keeps till June. Its great popularity is owing to its productiveness and long-keeping.

119. *Swaar.*—Large, pale lemon yellow, with dark dots, flesh tender, rich and spicy; tree a moderate grower, with dark shoots and large grey buds; with good culture it is one of the very best of apples.—November to May.

120. *Seek-no-further* (Westfield).—Medium to large, striped with dull red, and slightly russeted; flesh tender, rich and excellent; tree a good grower and bearer, and fruit always fair.—November to February.

121. *Spitzenburgh Esopus.*—Large, deep red, with grey spots, and delicately coated with bloom, flesh yellow, crisp, rich and excellent; tree rather a feeble, slow, spreading grower, and moderate bearer; esteemed in this State as one of the very best.—November to April.

Spitzenburgh Newtown.—From the little village of Newtown, Long Island; a handsome, round, dark red apple, marked with brown dots; flesh yellow, firm, mild and pleasant.—November to February. This was exhibited as " Vandervere" in nearly all the western collections at the Cincinnati Convention.

123. *Tallman Sweeting.*—Medium size, pale whitish yellow, slightly tinged with red; flesh firm, rich and very sweet; excellent for cooking; tree vigorous, upright and very productive.—November to April.

124. *Twenty Ounce* (Cayuga Red Streak).—A very large, showy, striped apple, of fair quality; tree an upright, compact grower, and fine bearer; excellent for baking, and of pleasant flavor, though not rich; one of the best very large apples, and popular in the markets.—October to January.

125. *Tewksbury Winter Blush.*—Small, yellow, with a red cheek, flesh firm, juicy, and fine flavored; a remarkably long keeper; tree a rapid, erect grower; suits the south best, as it requires a long season to manure it; origin, New Jersey.—January to July.

126. *Vandevere.*—Medium size, yellow, striped with red, and becoming deep crimson next the sun, flesh yellow, rich and fine; tree a fair grower and good bearer; succeeds best on light, warm, dry soils.—October to March.

127. *Wagener.*—Medium to large size, deep red in the

sun; flesh firm, sub-acid and excellent; tree very productive. A new and excellent variety, recently introduced from Penn Yan, Yates County, New York.—December to May.

128. *Winter Pearmain.*—Medium size, dull red stripes, on a yellowish ground, flesh tender, pleasant and aromatic; a moderate grower and bearer, best on warm soil. —November and March.

129. *Willow Twig.*—A western variety, particularly valuable for its productiveness and long keeping; shoots very slender, hence its name.

130. *Wells' Sweeting.*—A fine early winter sweet apple, from Newburgh, New York. Medium size, roundish, green, flesh tender, sprightly and rich. Tree a stout, upright grower and good bearer.—November to January.

131. *Wine Sap.*—Medium size, roundish, slightly conical, deep red; flesh very firm and crisp, with a sub-acid flavor. Tree very productive; fine for cider, for which it has been extensively grown.—November to April.

132. *Wine Apple* (Hay's Winter).—Large, roundish, slightly flattened, yellow striped, and clouded with bright red; flesh yellow, juicy, crisp and pleasant. A native of Delaware; succeeds well in many parts of the country.

133. *White Winter Calville.*—This is a celebrated French apple, and is one of the finest dessert varieties to be found in their markets at the present day. It is large, flat, ribbed, pale yellow, with a bright red cheek; flavor pleasant but not rich. Succeeds well in Canada.—November to March.

SMALL SELECT LISTS OF APPLES.

For Western New York the following are "unimpeachable:"

Nos. 1, 2, 4, 8, 9, 10, 14, 15, 23, 29, 38, 41, 44, 45, 47,

55, 72, 73, 76, 81, 83, 84, 87, 93, 97, 101, 105, 109, 113, 114, 119, 121, 127.

Twenty Choice Garden Varieties.—2, 9, 10, 15, 23, 29, 38, 41, 47, 50, 55, 84, 93, 97, 101, 105, 113, 119, 121, 127.

Twenty very large and beautiful sorts for Dwarfs.—2, 4, 23, 31, 35, 38, 27, 28, 41, 44, 45, 51, 59, 72, 73, 83, 101, 109, 127, 133.

Varieties that succeed well in the south and west.—1, 2, 4, 5, 6, 7, 8, 11, 12, 15, 23, 33, 38, 41, 47, 51, 61, 76, 79, 80, 82, 84, 86, 95, 102, 103, 107, 108, 109, 110, 111, 114, 116, 118, 126, 129, 132.

The "Western Horticultural Review" suggests the following for an orchard of one thousand trees at St. Louis.

Two hundred Rawle's Jannet; two hundred Pryor's Red; two hundred Newtown pippin; fifty golden russet (American, no doubt); thirty-five Newton Spitzenburgh; fifteen fall pippin; twenty-five each, yellow and white bellflower, early strawberry, early harvest, Benoni, Williams' favorite, Bohanan, and Gravenstein, Cooper, Rome Beauty, Rambo, Belmont, and Fameuse; one hundred Carthouse or Gilpin, Michael Henry and Smith's Cider; fifty "any others not rejected."

Samuel Walker, Esq., President of the Massachusetts Horticultural Society, one of the most experienced and critical judges of fruits in America, gives, in "Hovey's Magazine, vol. xv., 1849, the following list as being the best-suited to the meridian of New England, ranked according to merit thus: Nos. 114, 41, 72, 8 or 9, 113, 54, 99, 123, 4, 118, 81, 55, 88, 38, 84, 1, 3, 2, 39, 27, 94, 91, 104, 121.

SECTION 2.—SELECT PEARS.

CLASS I.—SUMMER PEARS.

1. *Amire Joannet* (St. John's).—Retained only on account of its earliness, and not recommended for small collections; the tree is a fine grower on both pear and quince, and a profuse bearer; fruit small, pale yellow, and sweet, soon mealy.—Twentieth to last of July. Foreign.

2. *Beau Present d'Artois.*—A good, juicy, half melting pear, medium size; tree very vigorous and productive on the quince.—September. Foreign.

3. *Bloodgood.*—An American pear of the first quality; medium size, buttery, melting, and rich; tree is a fair grower; succeeds well, and makes a fine pyramid on the quince.—August.

4. *Bartlett* (William's Bonchretien).—One of the most popular pears; large, buttery, and melting, with a rich musky flavor; tree a fair, erect grower; bears young and abundantly, both on pear and quince; the fruit ripens perfectly in the house, if gathered even two or three weeks before its time of ripening; this prolongs its season.—Last of September.

5. *Canandaigua.*—So named by the Congress Fruit Growers, in 1849; previously called "*Catharine.*" It is, in appearance, very similar to the Bartlett, nearly as large and as good; melting and fine flavored; tree is an upright, vigorous grower on both pear and quince, and very productive. In season middle of September.

6. *Dearborn's Seedling.*—Rather below medium size, pale yellow, melting and delicious; tree a rapid, fine grower, both on pear and quince, and bears young and profusely; one of the very best early summer pears, ripe

immediately after the Bloodgood.—About the last of August.

7. *Doyenne d'Été* (Summer Doyenne).—A beautiful little melting sweet pear; tree a fine grower and bearer, and succeeds well on the quince.—August.

8. *Edward's Henrietta.*—A medium-sized, greenish pear, melting, juicy, and sprightly; tree very productive; originated at New Haven, Conn.—September. Not much disseminated.

9. *Jargonelle, English.*—A good old variety, large size, flesh rather coarse, but juicy and pleasant; should be ripened in the house. The tree is a very strong grower, with purplish spreading shoots, and large foliage and flowers. It makes a fine prolific pyramid on the quince, but is not recommended for small collections.—Beginning of August.

10. *Leech's Kingsessing.*—A new pear recently introduced by Dr. Brinkle, of Philadelphia. It is described as being large, of a "sea-green" color, and "rich, buttery, and delicately flavored;" ripe last of August. It proves with us a strong grower, and will make a handsome pyramid on the quince.

11. *Madeleine.*—This is the earliest good pear we have; size medium, flesh melting, sweet and delicate; tree a fair grower, and good, early bearer on both pear and quince.

12. *Striped Madeleine*—is a variety, with wood and fruit striped with green, yellow and red, of same quality and season as the preceding.

13. *Muscadine.*—Medium size, melting, musky flavored and good. The tree has vigorous dark shoots, similar to Jargonelle; does not appear to succeed well on the quince.—Early in September.

14. *Moyamensing* (Smith's).—Originated in Philadelphia; large size, buttery, melting, and fine flavored; decays very soon after ripening; tree is a fair grower on

both pear and quince; ripens at Philadelphia from the middle of July till August; north proportionably later.

15. *Muscat Robert.*—A small, pale yellow, half-melting fruit, very juicy and sweet; tree hardy, productive, and vigorous, leaves large and flat, shoots quite yellowish; bears abundant crops on the quince; not so good as Madeleine or Bloodgood.

16. *Ott's Seedling.*—A small, yellow, delicious, high flavored pear; originated near Philadelphia, and brought to notice by Dr. Brinkle, who considers it nearly as good as the Seckel.—August and September.

17. *Osband's Summer.*—A medium sized, excellent fruit, half-melting, mild, and pleasant flavored; tree a fair, erect grower, on both pear and quince, and very productive; ripens here early part of August; origin, Wayne county, N. Y.; recently introduced.

18. *Rousselet Stuttgart.*—Medium size or rather small, greenish brown, half-melting, juicy, and fine flavored; tree a fine grower, and very productive on both pear and quince. German.—August and September.

19. *Rostiezer.*—Medium size, yellowish green, with a brown cheek; flesh juicy, sweet, and high flavored; the tree is vigorous, with dark-colored shoots; succeeds well on the quince; of German origin, and not long introduced here, but so far has proved first-rate in its season.—First of September.

20. *Summer Franc Real.*—Medium size, juicy, melting, and rich; tree very hardy, and a stout, fine grower and good bearer on both pear and quince; makes a beautiful pyramid; foliage and young shoots look mealy, being covered with a light down; fruit should always be ripened in the house.—End of August and beginning of September.

21. *Skinless* (Sanspeau of the French).—Rather small pyriform, pale yellow, with a tinge of red on the sunny

side; melting, juicy, and sweet; tree a rapid, erect grower, and very productive.—August.

22. *Tyson.*—Rather above medium size, melting, juicy, sweet, and fine flavored; tree very vigorous and rapid grower, both on quince and pear; one of the finest summer varieties; origin, Jenkintown, Pa.—September.

CLASS II.—AUTUMN PEARS.

23. *Andrews* (Amory or Gibson).—Rather above medium size, pyramidal, yellow, with a dull red cheek; juicy, melting, and good; tree a fair grower, and a good bearer; very popular in Boston. September and October. Native of Dorchester, Mass.

24. *Bleeker's Meadow.*—Medium size, round, half-melting; second rate; but the tree is exceeding hardy and vigorous, and bears enormous crops; fine for stewing, etc. —October.

25. *Buffum.*—Very similar to the White Doyenne, and nearly as good; buttery, sweet, and fine flavored; tree a remarkably vigorous, upright grower, with light reddish brown shoots; succeeds well on the quince.—Last of September.

26. *Beurre, Brown.*—A very fine old fruit; not well spoken of in New England, but does well in this State; fruit medium size to large, melting, very juicy, with a sprightly vinous flavor. It is rather a tardy bearer and poor grower on the pear, but on the quince it grows well and makes a fine prolific pyramid. It is inclined to overbear, and should be pruned close, to maintain its vigor.

27. *Beurre, Bosc.*—A large and beautiful pear, melting or nearly so, high flavored and delicious; a good grower on pear, but does not succeed on the quince, except double worked.—September and October.

28. *Beurre d'Amalis Panache.*—In quality similar to

Number 37, but the wood and fruit curiously striped with red and yellow, like the striped Madeleine and Doyenne.

29. *Beurre d'Anjou.*—A large fine pear, buttery and melting, with sprightly vinous flavor; tree a fine grower on both pear and quince.—October and November. Foreign; first imported by Col. Wilder.

30. *Beurre Capiaumont.*—Medium size, yellow, with a red cheek; melting, sweet, and rich, but somewhat variable; a good grower and abundant bearer; makes a beautiful productive pyramid on the quince.—October.

31. *Beurre Diel.*—One of the largest pears, buttery, rich, and fine; sometimes gritty at the core on pear stock; invariably first-rate on the quince; growth very strong and rapid, with large roundish leaves.—October and November; and if picked early and ripened gradually in the house, may be kept to December.

32. *Beurre, Golden of Bilboa.*—A large and beautiful pear of the first quality, buttery and melting, with a rich sprightly flavor; a strong, upright, handsome grower; succeeds particularly well on the quince.—September and October.

33. *Bergamotte Cadette* (Beurre Beauchamps).—A medium size, oval, melting rich pear; tree a fine grower, and very prolific both on pear and quince; makes a fine pyramid.—Ripe in November, and keeps well till December.

34. *Bergamot Gansels.*—Medium to large size, roundish, skin rough, brown; flesh melting, juicy, rich, and high flavored; rather a poor grower, with slender spreading branches and gray leaves; a tardy bearer, does not succeed on the quince.—October.

35. *Bezi de Montigny.*—A very fair, medium sized fruit, melting and sweet, with a pleasant, musky flavor; very vigorous and productive; makes a fine fruitful pyra-

mid on the quince, nearly as good as White Doyenne.—October.

36. *Belle Lucrative* (Fondante d'Automne).—Medium size, melting and delicious; a fair, upright grower; makes a beautiful pyramid on the quince, and bears early and abundantly; first quality in all respects.—September and October.

37. *Beurre d'Amalis.*—A large, melting, fine pear, resembling the Brown Beurre, and with the same high vinous flavor; a strong grower, and most abundant bearer on the quince.—September and October.

38. *Comte de Lamy.*—A medium-sized, melting, fine flavored fruit; a good grower; makes a fine pyramid on the quince.—October.

39. *Cushing.*—Medium size, oblong, pale green, brownish next the sun; juicy, melting, and delicious; tree very productive; one of the very best Massachusetts varieties. —September.

40. *Dunmore* (Knight's).—A large, fine pear, with a sprightly sub-acid flavor like the Brown Beurre; rather variable; sometimes first-rate—September.

41. *Dix.*—A large, fine pear, melting, juicy, and rich; a fair grower; slender, yellowish shoots; succeeds on the quince double worked, but not otherwise.—October. Origin, Boston.

42 *Duchesse d'Angouleme.*—The largest of all our good pears; it attains its highest perfection on the quince, and is a beautiful, vigorous tree; profitable for market.— October and November.

43. *Duchesse d'Orleans.*—A new, large, and delicious pear, "with the flavor of Gansell's Bergamot;" succeeds well on the quince, and bears quite young; has proved fine at Boston, but not so good with us.—October.

44. *Doyenne Boussoch.*—A new, large, delicious pear,

like a very large White Doyenne; tree a strong, rapid grower; succeeds well on the quince.—October.

45. *Doyenne White.*—A well known and universally esteemed variety of the highest excellence; growing and bearing equally well on both pear and quince; young trees on the quince are inclined to bear too much, on this account the fruit requires thinning, and the tree pretty close pruning, to keep up a vigorous growth.—October and November.

46. *Doyenne Gray.*—Similar in quality to the preceding; fruit of a reddish russet color; tree not quite so strong a grower.

47. *Doyenne Panache* (Striped Doyenne).—Similar in character to the White; wood and fruit are curiously striped.

48. *Excellentissima.*—A very beautiful and excellent new Belgian variety, as large as the Bartlett; oblong, pyriform, yellow, with a tinge of red in the sun; buttery, melting, and rich.—October. It must become a most popular variety.

49. *Forelle or Trout Pear.*—A beautiful German pear, finely speckled, buttery, melting and rich; succeeds well on the quince; bears early and profusely.—November.

50. *Flemish Beauty.*—A large, beautiful, melting, rich pear; tree vigorous and fruitful: has not succeeded well on the quince with us yet.—September and October.

51. *Frederick of Wurtemberg.*—A large and beautiful pear, sometimes first-rate, and often insipid; a vigorous grower both on pear and quince, and an early good bearer. —September.

52. *Fulton.*—A native pear, round and russety, melting, rich and excellent, hardy and vigorous; succeeds well in the north where many others are tender.—October and November.

53. *Henry IV.* (We have sent out some trees of this

variety, imported as "Ananas").—A medium sized, melting, delicious pear, of a dull greenish color; a free stout grower, and a most profuse bearer; succeeds well on the quince; should be ripened in the house.—September.

54. *Heathcote.*—A buttery, melting, fine-flavored pear, about the size of White Doyenne, hardy and productive. —September. Originated in the vicinity of Boston.

55. *Howell.*—A very good and very handsome variety, originated at New Haven, Conn. Large or medium, obovate, inclining to pyramidal, lemon yellow tinged, with red in the sun; melting, juicy and vinous, rather coarse Tree vigorous, branches rather spreading and drooping.— September and October.

56. *Hanners* (Hannas).—A native of Boston or vicinity; of medium size, resembling the Cushing, with somewhat the flavor of White Doyenne.—September.

57. *Henkel.*—Medium size, roundish obovate, yellow. slightly russeted; buttery, melting and fine. One of the very best Belgian varieties.—November and December.

58. *Hacon's Incomparable.*—A very large round pear, buttery, melting and rich. Trees spreading and very productive. One of the finest English varieties, but a little variable in this country.—October.

59. *Harvard.*—A very popular and profitable market fruit around Boston; large, oblong, russety, melting and juicy; should be house ripened. The tree is vigorous and upright; a tardy but very abundant bearer.—September.

60. *Johonnot.*—Medium size, roundish obovate, greenish yellow, slightly russeted, rather coarse but melting, rich and musky. The tree is a good grower, succeeds well on the quince. Originated in Salem, Mass.

61. *Knight's Seedling.*—Originated in Rhode Island. Large, melting, sweet and good; should be gathered before ripe; tree a good grower.—September to October.

62. *Louise Bonne de Jersey.*—One of the finest of all

pears, large, beautiful and delicious; it succeeds well both on pear and quince, but on the latter, especially, it is al. that can be desired.—September and October. Foreign.

63. *Long Green* (Verte Longue).—Really long and green, juicy, sweet and good; a fine strong grower and good bearer on the quince.—October. Foreign.

64. *Long Green, Striped* (Verte Longue Panache).—A variety of the preceding, with striped wood and fruit, but nferior in quality.

65. *Las Canas.*—A fine Belgian variety, first fruited by Mr. Manning. Medium size, pyriform, yellow and slightly russeted, juicy, melting and fine.—October.

66. *Lodge.*—Medium size, brownish, russety, juicy, melting, rich; native of Philadelphia; usually first-rate. —September and October.

67. *Marie Louise.*—A large melting pear of the first quality; the tree is a straggling, crooked grower, but hardy, and bears young and abundantly; does not succeed on the quince.

68. *Napoleon.*—A large, juicy, melting, fine fruit; tree vigorous, hardy and productive, makes a fine pyramid on the quince; bears young, ripens in November, and may be kept till December. Should be ripened in a warm room.

69. *Onondaga.*—See Swan's Orange.

70. *Oswego Beurre.*—An excellent new pear, from Oswego, New York, medium size, melting, with a rich vinous flavor. Tree very hardy and productive; makes a beautiful pyramid on the quince; a most valuable pear; ripens in October to December.

71. *Paradise d'Automne.*—A large, fine melting pear, of the first quality; resembles Beurre Bosc in shape and color.—October.

72. *Pratt.*—A native of Rhode Island, medium size,

melting, buttery, rich and good; nearly first rate.—September and October.

73. *Petre.*—Native of Pennsylvania; medium size, obovate, pale yellow, slightly russeted, buttery, melting and rich; generally first rate.—September.

74. *Pennsylvania.*—Medium to large size, half melting highly esteemed at Philadelphia, where it originated; does not prove so good in other places. Tree a good grower, both on pear and quince.—September.

75. *Surpass Virgoulouse* (or Virgalieu). A very fine fruit, nearly equal in all respects to the White Doyenne, introduced by the late Mr. Parmentier, of Brooklyn.—October.

76. *Seckel.*—The highest flavored pear known; considered as the standard of excellence; a stout, erect grower, not rapid, a good bearer; grows well on the quince with us.—September and October.

77. *St. Ghislain.*—A medium sized, fine melting pear; tree a rapid and beautiful grower and good bearer; should be ripened in the house.—September.

78. *Stevens' Genesee.*—A large, roundish, buttery, fine-flavored pear, vigorous and highly productive; succeeds well on both pear and quince; a native of Monroe county, New York.—September and October.

79. *Swan's Orange* (Onondaga).—A very large, melting, high-flavored pear, vigorous and extremely productive, one of the best pears of its size and season; succeeds well on the quince.—October and November.

80. *Urbaniste* (Beurre Picquery of the French).—A large, melting, buttery pear, a tardy bearer on the pear, but succeeds well on the quince.—October and November.

81. *Van Mons Leon Leclerc.*—A very large pear, four and a half inches in length, and three in diameter, of an orange color at maturity. Tree vigorous and productive, succeeds well on the quince, and bears quite young;

cracks a little in some seasons, and should have a rich, warm soil.—October and November.

82. *Washington.*—A medium sized, beautiful pear, sweet and delicious; a fine grower on the pear, but does not succeed on the quince.—Middle of September.

83. *Wilkinson.*—A very hardy and productive Rhode Island variety of second quality, medium size, obovate, yellow, melting, sweet and rich. Tree upright and vigorous.—October to November.

84. *Wilbur.*—Medium, obovate, greenish and russety; rather coarse, melting, juicy and good. Native.

CLASS III.—WINTER PEARS.

85. *Beurre d'Aremberg.*—One of the finest winter pears, large, melting, rich, vinous flavored, ripens well without any extra care; tree vigorous and productive; succeeds well on the quince.—December to January.

86. *Beurre, Easter.*—A very large, fine melting pear; better on the quince than on the pear; keeps till spring.

87. *Beurre Gris d'Hiver Nouveau.*—A large, new, melting, buttery pear, of the highest quality, flavor rich and vinous, like the brown Beurre, but milder; tree succeeds well on the quince, bearing quite young.—November and December, and may be kept till January.

88. *Beurre Rance.*—A fine, melting, rich pear, keeping till spring; tree a poor grower.

89. *Chaumontel* (English):—A large, fine, buttery, melting, rich pear; should have a warm soil and situation; succeeds well on the quince.—December.

90. *Columbia.*—A large, handsome, native pear, melting, buttery and rich; tree vigorous and remarkably productive; grows well on the quince with us.—November and January.

91. *Doyenne d'Hiver Nouveau or d'Alençon.*—A large

and very fine late-keeping variety received from France; golden yellow, with a brown tinge in the sun, melting, buttery and rich; tree upright and vigorous, and very productive on the quince; have had it very fine on the 1st of March, ripened in the cellar; will prove very valuable.

92. *Glout Morceau.*—A large, melting, buttery, sweet pear; tree vigorous and productive; like the Duchesse d'Angouleme, Louise Bonne, and some others, it is decidedly superior on the quince, and makes a beautiful pyramid.—December.

93. *Josephine de Malines.*—A new Flemish winter pear, pronounced both in France and England to be the finest winter variety; medium in size, melting and rich; the tree is a moderate grower, with quite small leaves; it appears to succeed well on the quince; keeps till spring.

94. *Lawrence.*—A fine, large, melting, rich flavored pear, a native of Long Island; tree a fair grower on both pear and quince; a regular and abundant bearer.—November to February. Ripens well in the cellar.

95. *Passe Colmar.*—Large, buttery, and rich; tree a fine, free grower on both pear and quince; so disposed to over-fruitfulness that thinning is quite necessary to obtain fine fruit.—December.

96. *Sieulle* (Doyenne Sieulle).—A large, roundish, melting pear, that keeps till January; tree vigorous and upright; succeeds remarkably well on the quince.

97. *St. Germain, Prince's.*—A medium sized, juicy, fine flavored fruit; ripens in the cellar, like an apple through the winter; a moderate grower and good bearer.—March.

98. *St. Germain.*—An old variety, large, melting, and sweet; succeeds well on the quince; bears young and abundantly.—December to January.

99. *Vicar of Winkfield or Monsieur Le Curé.*—A

large, long pear, fair and handsome, of good, but not first-rate quality; tree a most vigorous grower on both pear and quince, and on the latter makes a beautiful and productive pyramid; one of the most valuable of all late pears.—November to January. Ripens well in the cellar

100. *Winter Nelis* (Bonne de Maline and Beurre de Maline, of some French catalogues).—One of the best of early winter pears, medium size, melting and buttery, with a rich, sprightly flavor; tree is rather slender, straggling growth.—November and January.

CLASS IV.—SELECT BAKING AND STEWING PEARS.

101. *Bonchretien* (Flemish).—Medium to large size; tree vigorous, spreading, and irregular, and bears great crops; keeps through winter.

102. *Bonchretien* (Spanish).—Medium to large, pyriform, tapering to the stalk, yellow with a red cheek; cooks well.—December and January.

103. *Cattillac.*—Very large, roundish, bears quite young and abundantly on the quince; keeps all winter.

104. *Chaptal.*—Very large, somewhat resembling Duchesse d'Angouleme; keeps till spring; cooks finely, and is sometimes tolerable for eating; tree vigorous; bears very young on the quince.

105. *Easter Bergamot.*—Medium size, rough, greenish; keeps well, and cooks finely; trees remarkably vigorous, erect, and fruitful; bears quite young on the quince.

106. *Pound* (Angora, Uvedale's St. Germain).—Monstrous size; often weighs two pounds; stews well; tender, and of a rich crimson color; trees vigorous, and very productive; liable to be blown off standard trees; succeeds well on the quince, and bears young; keeps all winter.

SELECT ASSORTMENTS OF PEARS.

Profitable varieties for market orchards.—Nos. 4, 31, 42, 45, 46, 50, 62, 78, 85, 86, 94, 99, 106.

Ten very hardy prolific sorts.—Nos. 20, 24, 25, 52, 59, 70, 94, 97, 99, 106.

The great market pear of Western New York, is the *White Doyenne* or *Virgalieu*. No better can be desired of the season; for summer the *Bartlett*. The *Windsor* or *Summer Belle*, and the *Summer Bonchretien*, are both very profitable pears, at present brought into our markets in large quantities. Neither of them is described, because it is not desirable to extend their cultivation whilst we have the *Bartlett*.

Twenty-five fine varieties for the garden, on quince stocks.—Nos. 3, 4, 6, 19, 22, 31, 32, 33, 36, 37, 42, 45, 46, 53, 62, 68, 76, 78, 79, 85, 86, 87, 92, 99, 106.

Any or all of these may be chosen without running any risk of a failure.

First-rate sorts for pear stocks, or to be double worked on the quince.—Nos. 27, 34, 41, 50, 67, 71.

NEW AND RARE PEARS, RECENTLY INTRODUCED, THAT GIVE PROMISE OF EXCELLENCE.

THOSE THAT HAVE BEEN PROVED TO SUCCEED ON THE QUINCE ARE DESIGNATED BY A (q).

107. *Adele de St. Denis.*—New Belgian, medium size, very handsome, russety, melting, vinous, and perfumed.—October.

108. *Arch Duc Charles.*—Medium size, melting; tree a vigorous, fine grower, both on pear and quince.—October.

109. *Alpha* (New Belgian).—Medium size, greenish yellow, buttery and fine; highly recommended by Mr. Manning.—October.

110. *Arbre Courbe or Amiral* (Van Mons).—Medium to large, melting and delicious; branches irregular and crooked.—September.

111. *Belle et Bonne des Zees or Bonne des Zees.*—A new, large, and fine Belgian variety, ripening immediately after the Bartlett; obovate, yellow, red next the sun; melting and perfumed.

112. *Brandywine.*—Recently introduced by Dr. Brinkle, of Philadelphia; originated in that vicinity; ripens same season as the Bartlett; obovate, yellow, and slightly russeted; melting, sweet, and perfumed.—September.

113. *Beurre Benoist.*—Large, obovate, yellow with a red cheek; melting, juicy, and sweet.—October. From France recently; found in a hedge.

114. *Beurre Bretonneau* (Esperin, Belgium).—Medium to large, oval, slightly pyramidal; half melting; tree vigorous and productive; keeps till April or May.

115. *Beurre Clairgeau.*—A new French variety, described to us as being as large as Duchesse d'Angouleme, and of excellent quality.—October and November.

116. *Bezi Sans Pareil.*—A new winter variety from France; large, greenish, half melting.—February.

117. *Beurre Curtet* (Bouvier).—Medium size, melting, and fine (q).—October. French.

118. *Beurre Davis.*—Large, melting, productive (q).—October. French.

119. *Beurre Duval.*—Large, melting; tree vigorous and productive (q).—October and November. French.

120. *Beurre Giffard or Giffart.*—Medium size, melting; tree has erect, slender branches; productive; one of the best new, early varieties (q).—August. French.

121. *Beurre Goubault.*—Medium size, roundish, half melting; tree vigorous and very prolific; bears quite young (q); new.—September. Angers, France.

122. *Beurre Moire.*—Medium size, melting; tree vigo-

rous and productive (q).—September and October. French.

123. *Beurre Superfine.*—Medium size, melting, very productive; new (q).—October. French.

124. *Beurre St. Nicholas.*—Large, green, and russet, melting, a little coarse, juicy, and high flavored.—September and October. French.

125. *Beurre Hardy.*—A new variety introduced by Jamin, of Paris; large, melting, and good; tree a vigorous and beautiful grower on the quince.—October.

126. *Beurre de Waterloo.*—Medium, obovate pyriform, greenish yellow, with a blush next the sun; flesh buttery, melting, and high flavored; received from France in 1843; was the best pear we tasted in 1850; ripe latter end of October.

127. *Beurre Langelier.*—A splendid new variety, introduced by Mr. Langelier of the isle of Jersey. Mr. Hovey, who has had it bear, gives it the highest character, and thinks it will prove one of the finest winter varieties from abroad. Large, greenish, with a shade of red in the sun; melting, juicy, and vinous; tree a beautiful grower, with large, shining foliage; succeeds finely on the quince.—December and January.

128. *Brandes St. Germain* (Van Mons).—Medium, pyriform, oblong, green and russet, melting, sugary, and rich; tree vigorous; shoots slender and spreading.—December to March. Will prove a valuable winter variety; first introduced by Mr. Manning.

129. *Broompark* (Knight's) English.—Medium, roundish, of a beautiful cinnamon russet, melting and juicy, "partaking of the flavor of a melon and pine-apple."—January.

130. *Burlinghame.*—A seedling raised in Marietta, Ohio; medium size, very productive, melting, and good. —July and August.

131. *Colmar Musqué.*—Medium size, turbinate, golden yellow, texture and flavor of Bartlett; ripe in October; a most delicious variety; bears young and abundantly; received from France in 1848.

132. *Cabot* (Massachusetts).—Medium size, buttery and sweet, very productive.—September and October.

133. *Catinka* (Esperin).—Large, melting, and excellent; new; said to be in eating for five or six weeks (q).

134. *Colmar Bonnet* (Van Mons).—Medium size, melting and sugary (q).—September and October.

135. *Colmar d'Aremberg.*—Very large, of second quality; tree very vigorous and productive (q).—October and November.

136. *Commodore.*—Medium size, buttery, and sweet.—October.

137. *Chapman* (Penn.).—Large, half melting.—September and October.

138. *Chancellor* (Penn.).—Large, melting, and rich; new.—September and October.

139. *Delices de Jodoigne* (Bouvier).—Large, melting; tree vigorous and productive; new (q).—November.

140. *De Bavay.*—Large, melting, said to be first quality (q).—September.

141. *Duc de Bordeaux* (Epine Dumas).—Medium size, half melting, vigorous, and productive (q).—November.

142. *De Lepine.*—Medium size, half melting, productive (q).—September.

143. *Delices d'Hardempont.*—Medium size to large, melting, productive (q).—November and December.

144. *Dillen* or "*Doyenne Dillen.*"—Medium size, melting and good (q).—December.

145. *Doyenne Goubault.*—Medium size, melting and good, very productive (q).—Winter.

146. *Doyenne Robin.*—A new variety from Angers,

France, said to be very large, beautiful, and excellent; ripens in October.

147. *Doyenne Rose.*—One of the most beautiful of all pears, resembling the White Doyenne in size and form, but not so good; Same season.

148. *Duchesse de Mars.*—Medium, roundish, obovate, pale yellow, fair and smooth, juicy, melting, and very highly perfumed.—October. Received from France in 1848; succeeds on the quince; tree rather delicate.

149. *Eyewood* (Knight, Eng.).—Medium size, melting and rich, high flavored.—November.

150. *Ferdinand de Meester, or Rousselet de Meester* (Van Mons).—Medium size, melting and good.—September and October.

151. *Fleur de Niege.*—Medium size, melting and productive (q).—October.

152. *Fortunee.*—Rather small, melting, high flavored; succeeds well on the quince; keeps till spring. Episcopal has proved synonymous with this.

153. *Fredrika Bremer.*—A variety recently brought to notice by Mr. John C. Hastings, of Clinton, N. Y. It is large, obovate inclining to pyriform; green, changing to yellow as it matures; melting, buttery, and sprightly; may prove to be a fine variety. October and November. The tree is said to be very productive, and some specimens to attain the weight of sixteen ounces.

154. *Fondante de Maline.*—A new Belgian winter variety, melting and good; very productive; succeeds well on the quince. Keeps till February.

155. *Figue.*—A very distinct, greenish pear; medium size, pyriform, stem fleshy; melting, juicy, and good. November. Tree vigorous, and exceedingly productive on the quince.

156. *Gratioli of Jersey.*—Medium size, melting; very sweet and good; succeeds well on the quince.—October.

157. *Hull* (Mass.).—Medium size, melting, fair and good.—September and October.

158. *Inconnue Van Mons.*—Medium size, melting, and fine.—January to February.

159. *Jalousie de Fontenay Vendee.*—Medium size, melting and rich; new.—September. Has proved excellent so far.

160. *Jones's Seedling* (Phila.)—New, and said to be excellent; grows well on the quince.

161. *Knight's March Bergamot.*—One of Knight's best seedlings, described as resembling the Autumn Bergamot; buttery and rich; valuable for its long keeping.—March. Very hardy and productive.

162. *Kirtland.*—A seckel seedling, raised in Ohio by H. P. Kirtland, Esq., and introduced by Prof. Kirtland, of Cleveland, who describes it as medium size, globular ovate, crimson russet, varying to a dull green; melting, juicy, rich, and in the highest degree delicious; tree has the thrifty habit of White Doyenne.—September.

163. *Louise de Boulogne.*—Large, breaking, keeps through winter; succeeds on the quince.

164. *Louise d'Orleans* (Van Mons).—Medium size, oblong, brownish green; melting and sugary.—November.

165. *Moccas* (Knight's, Eng.).—Medium size, obovate, brown; melting, juicy, and high flavored; tree very hardy and productive.—December.

166. *Monarch* (Knight's).—Spurious varieties have been disseminated. The true one is large, roundish obovate, brownish, buttery, and slightly musky; tree hardy and productive, but a tardy bearer; succeeds double worked on the quince.—January.

167. *Muskingum.*—A native of Ohio; rather large, roundish, greenish yellow, russeted, melting, juicy, sweet, and high flavored. September. It is said to be hardy, productive, and a fine grower; may prove valuable.

168. *McLaughlin.*—A native of Maine; medium size, obovate, brownish yellow; a little coarse, but juicy and rich; very hardy and productive; may be valuable for the North.—November and December.

169. *Nouveau Poiteau* (Van Mons).—A large, fine, melting pear; has fruited at Boston, and is pronounced excellent.—October and November. Tree vigorous and productive.

170. *Osborne.*—Medium to small, bright yellow, melting, juicy, and sweet; tree vigorous; originated in Indiana, proves good at Cincinnati; may prove a valuable early variety for the West.—August at Cincinnati.

171. *Passe Tardive.*—Large, breaking, productive; for cooking all through winter.

172. *Queen of the Low Countries.*—Medium to large, half-melting.—October and November.

173. *Reine d'Hiver.*—Medium, half-melting; productive.—December to January.

174. *Seigneur d'Esperin.*—Medium size, melting; first quality.—October.

175. *St. Andre.*—Medium size, half-melting, very productive.—October and November.

176. *Sageret.*—Medium size, melting, sweet and sugary.—December to March.

177. *St. Michael Archangel.*—Large, melting, very productive.—October.

178. *Suzette de Bavay.*—Medium, melting, first quality, remarkably productive; is said to keep all winter; best in March and April.

179. *St. Dorothée.*—Large to medium, greenish yellow, russeted; melting, sprightly, and fine. October. This has borne with Mr. Hovey, who gives it a very high character.

180. *Triomphe de Jodoigne* (Bouvier).—Very large,

melting; tree very vigorous and productive. November and December. Has proved good at Boston.

181. *Tarquin.*—Large, coarse; for cooking only; tree very vigorous and productive; is said to keep two years.

182. *Viscomte Spoelberg.*—Medium size to small, yellow tinged with red next the sun; buttery and melting; first-rate under good culture; succeeds well on the quince.—November.

Section 3.—Quinces.

1. *Apple-Shaped or Orange.*—Large, roundish, with a short neck; of a bright golden yellow color; tree has rather slender shoots and oval leaves; very productive. This is the variety most extensively cultivated for the fruit.—Ripe in October.

2. *Pear-Shaped.*—This has generally more of a pyriform shape than the preceding; the fruit is larger and finer, the tree stronger.

3. *Portugal.*—The fruit of this is more oblong than the preceding, of a lighter color and better quality, but not so good a bearer; the shoots are stouter, and the leaves thicker and broader; usually propagated by budding or grafting on the Apple Quince. A week or two later than the Apple.

4. *Angers.*—A variety of the Portugal, the strongest grower of all the quinces, and the best for pear stocks. The fruit is also said to be larger and rather better than any of the others. We have not seen it yet, but expect our trees to bear this season, 1851.

5. *Upright.*—A variety with slender erect branches; grows more freely from cuttings than any other. We have not fruited it, nor found it anywhere described, but have trees now showing fruit buds. Received among stocks from France.

6. *Chinese.*—Usually cultivated for ornament. Quite different in appearance from the others. The leaves are glossy, sharply and beautifully toothed; the fruit is large, oblong, bright yellow, and keeps till spring; little used. The flowers are large and showy, with the fragrance of the violet; worked on the other sorts; rather tender, requiring a sheltered situation. A very tardy bearer.

7. *Japan.*—This is very distinct from all the others; very bushy, thorny, and hardy. There are two varieties. The common one has beautiful bright red blossoms, and the other blush; the most beautiful of all our hardy spring flowering shrubs. Fruit about as large as a chicken's egg; green, and quite unfit for use.

SECOND DIVISION.—STONE FRUITS—APRICOTS, CHERRIES, PEACHES, NECTARINES AND PLUMS.

SECTION 4.—SELECT APRICOTS.

1. *Breda.*—Small, round, dull orange, marked with red in the sun, flesh orange colored, juicy, rich and vinous; parts from the stone, kernel sweet, tree hardy, robust and prolific.—End of July and beginning of August.

2. *Early Golden* (Dubois).—Small, pale orange, flesh orange, juicy and sweet; kernel sweet; tree very hardy and productive. The original tree at Fishkill is said to have yielded $90 worth of fruit in one season.—Beginning of July.

3. *Large Early.*—Large, orange, with a red cheek, flesh sweet, rich and excellent, parts from the stone; tree vigorous and productive.—Beginning of August.

4. *Moorpark.*—One of the largest and finest apricots, yellow, with a red cheek, flesh orange, sweet, juicy and rich, parts from the stone; growth rather slow, but stout and short jointed; very productive.

5. *Orange.*—Medium size, orange, with a ruddy cheek, flesh rather dry, requires ripening in the house; adheres slightly to the stone.—End of July.

6. *Peach.*—A very large, handsome and excellent variety, quite similar to the Moorpark; the shoots are not so short jointed, and the fruit a degree larger.

7.—*Purple or Black Apricot.*—This is quite distinct in all respects from others, very much like a plum, small, pale red, purple in the sun, flesh yellow, juicy and pleasant. The tree has slender dark shoots, and small, oval, glossy foliage. It is as hardy as a plum, and therefore worthy of attention where the finer sorts are too tender. —August.

Nos. 1 and 2 are the surest and most abundant bearers, but 3, 4 and 6 are the largest and finest. No. 7 is only recommended by its hardiness, for localities where the others do not succeed.

Section 5.—Select Cherries.

Class I.—Heart Cherries.

Fruit heart shaped, with tender sweet flesh. Trees of rapid growth, with large, soft drooping leaves.

1. *American Amber.*—Medium size, amber, shaded and mottled with bright red; tender, juicy, sweet and delicious; hangs very long on the tree without rotting; remarkably vigorous and productive.—End of June till middle of July.

2. *Bauman's May.*—Small, dark red; tender, juicy and sweet. Tree a vigorous grower, and a most abundant bearer. Ripens very early; middle of June here. French

3. *Black Heart.*—An excellent old variety; rather large, black, tender, juicy and rich. Tree grows large, and is very prolific.—Beginning of July. French.

4. *Black Eagle.*—Large, black, tender, juicy, rich and high flavored. Tree a rapid, stout grower and productive. Ripe beginning of July to the 15th. English.

5. *Black Tartarian.*—Very large, purplish black, half tender; flavor mild and pleasant. Tree a remarkably vigorous, erect and beautiful grower, and an immense bearer.—Ripe last of June and beginning of July. One of the most popular varieties in all parts of the country. Russian.

6. *Burr's Seedling.*—Large, pointed; flesh color in the shade, pale red in the sun; tender, sweet and delicious. In luxuriant foliage and stateliness of growth it surpasses even the Black Tartarian.—Beginning of July. New; origin, Perrinton, Monroe county. New York.

7. *Coe's Transparent.*—Medium size, pale amber, red and mottled next the sun; tender, sweet and fine.—End of June here. Tree vigorous and erect. Origin, Middletown, Conn.

8. *Davenport's Early.*—Very similar in all respects to Black Heart, but a few days earlier. American.

9. *Downer's Late Red.*—Rather large, light red, tender and juicy; slightly bitter until fully ripe, when it is most delicious. Tree is a vigorous erect grower, and productive.—Tenth to twentieth of July. American, and one of the best of all.

10. *Early White Heart.*—Medium size, yellowish white, red in the sun; tender and sweet, growth moderately vigorous and erect.—Middle and last of June.

11. *Early Purple Guigne.*—Small to medium size, purple, tender, juicy and sweet. Growth slender and spreading.—Ripe at same time as Bauman's May. French.

12. *Elton.*—Large, pointed; pale yellow, nearly covered with light red; half tender, juicy, rich and delicious. Tree vigorous, spreading and irregular.—End of June. English.

13. *Knight's Early Black.*—Large, black, tender, juicy, rich and excellent. Tree vigorous and very productive; branches spreading.—Ripe a few days before Black Tartarian. English.

14. *Manning's Mottled.*—Rather large, amber shaded and mottled distinctly with red; tender, sweet and delicious. Tree erect, vigorous and productive.—End of June. Massachusetts.

15. *Sweet Montmorency.*—Small, light red, tender and sweet. Tree vigorous, erect and productive.—Ripens about the same time as Sparhawk's Honey, or a few days later. American.

16. *White French Guigne* (probably the "Merisier a gros fruit blanc," of the French).—A distinct and beautiful cherry, rather large, creamy white, flesh tender and melting; juice colorless, sweet, with a scarcely perceptible degree of bitterness; not attacked by the birds, like red and black cherries. Tree is vigorous and very productive.—Middle of July. French.

17. *Wilkinson.*—Medium size, black, tender, juicy and rich. Tree vigorous, erect and productive.—Ripens late, succeeds Downer's. Massachusetts.

18. *Sparhawk's Honey.* Medium size, roundish, light red, sweet and delicious; stone large. Tree a vigorous, pyramidal grower and very productive.—Ripens with Downer's late, and hangs long on the tree; a great favorite with most people. Massachusetts.

CLASS II.—BIGARREAU CHERRIES.

These are chiefly distinguished from the preceding class by their firmer flesh. Their growth is vigorous, branches spreading, and foliage luxuriant, soft and drooping.

19. *Bigarreau, or Yellow Spanish*—Large, pale yel-

ow, with a bright red cheek in the sun; flesh firm, juicy and delicious; one of the best, most beautiful, and popular of all light colored cherries. Tree vigorous and productive.—End of June. Turkish.

20. *Buttner's Yellow.*—Medium size, yellow, flesh crisp, juicy and sweet. Tree vigorous and productive. Its peculiar and beautiful color makes this sort desirable.—End of July.

21. *China Bigarreau.*—Medium size, oval, red, beautifully speckled; firm, sweet and rich, with a scarcely perceptible bitterness. Tree vigorous, erect, and a most profuse bearer; a very distinct and pretty variety.—Beginning of July. Hangs long on the tree.

22. *Flesh-Colored Bigarreau* (Bigarrean couleur de chair).—A large and beautiful cherry, resembling the Elton, and ripening about the same time. French.

23. *Florence.*—A beautiful cherry, resembling the Bigarreau; but firmer, and a week later. From Florence.

24. *Gridley or Apple Cherry.*—Medium size, dark brown, nearly black; flesh very firm, sprightly sub-acid, high flavored. Tree grows rapidly and erect, and bears immense crops. Its firmness and lateness make it very valuable for market.—Middle to last of July. Mass.

25. *Hildesheim Bigarreau.*—Medium size, yellow, red in the sun; flesh firm, sweet and agreeable. Tree is a good grower, but the ends of the young shoots are apt to get winter killed here.—Beginning of August. German.

26. *Holland Bigarreau.*—A very large and beautiful cherry; pale yellow, covered with bright red in the sun; flesh firm, juicy, sweet and fine flavored. Tree vigorous, with spreading, irregular branches.—End of June and beginning of July. Dutch.

27. *Large Heart-shaped Bigarreau* (Gros Couret).— Large, dark, shining brown; firm, rich and excellent

Tree vigorous, branches spreading.—Middle of July. French.

28. *Madison Bigarreau.*—Medium size, amber, covered with red in the sun; flesh half tender, sweet and fine flavored.—End of June and beginning of July. American.

29. *Merveille de Sept.*—A new French cherry, remarkable *only* for its lateness.—Ripens with us the last of August. Tree a vigorous grower and good bearer. Fruit small, firm, rather dry and sweet.

30. *Napoleon Bigarreau.*—A magnificent, large cherry, surpassing in size and beauty all the others; pale yellow, with a bright red cheek; flesh very firm until fully ripe, when it becomes tender, juicy and sweet. Tree is a vigorous grower, and bears enormous crops.—Beginning of July. French.

31. *Rockport Bigarreau* (Dr. Kirkland).—Large, pale amber in the shade, light red in the sun; half tender, sweet and good. Tree vigorous and erect.—Ripe same time as Black Tartarian. Ohio.

32. *Tradescant's Black Heart* (Elkhorn).—Very large, black, firm, juicy and good. Tree vigorous and upright, with peculiar gray bark. A great bearer, and so late as to be very valuable.—Middle and last of July. England.

33. *Tardive d'Argental.*—Large, long, dark, shining, red, nearly black; tender, when ripe; juicy, with a peculiar flavor, something like raspberry. Tree is an upright, vigorous grower, with peculiar small, light, wavy leaves. —Middle of July.

CLASS III.—DUKE AND MORELLO CHERRIES.

These two classes of cherries are very distinct from the preceding. The trees are of smaller size and grow slowly; the leaves are thicker and more erect, and of a deeper green. The fruit is grenerally round, and in color varying from light red, like *Belle de Choisy*, to dark brown, like *Mayduke* or *Morello*.

The *Dukes* have stout, erect branches usually, and some of them, like *Belle de Choisy* and *Reine Hortense*, quite sweet, whilst the *Morellos* have slender, spreading branches, and acid fruit invariably. These two classes are peculiarly appropriate for dwarfs and pyramids, on the mahaleb stock, and their hardiness renders them well worthy of attention in localities where the *Hearts and Bigarreaus* are too tender.

34. *Belle de Choisy.*—Medium size, amber shaded and mottled with red; tender, melting, sweet and rich; rather a shy bearer; tree makes a pretty pyramid.—End of June. French.

35. *Belle Magnifique.*—A magnificent, large, red, late cherry; excellent for cooking, and fine for table when fully ripe, rather acid, tender, juicy, rich; tree is a slow grower, but a most profuse bearer; makes a fine dwarf or a pyramid on the mahaleb.—Last of July. French.

36. *Carnation.*—Large, light, red mottled with orange; tender, juicy, a little acid, rich, and excellent; tree is a good grower and a profuse bearer; makes a fine dwarf.—Middle and last of July.

37. *Donna Maria.*—Medium size, dark red, tender, juicy, acid, rich, fine for cooking; tree small, very prolific.—Middle of July. French.

38. *Du Nord Nouvelle.*—A new French morello, ripens all through August; medium size, bright red, tender, acid; useful on account of its lateness; makes a beautiful dwarf or pyramid.

39. *Early Richmond, Kentish* or *Montmorency.*—An early, red, acid cherry, very valuable for cooking early in the season.—Ripens through June.

40. *Flemish Montmorency.*—A remarkably short-stem-

med, flattened cherry; medium size, red, tender, juicy, acid, good for cooking; rather a poor bearer, but curious.

41. *Indulle, Nain Precoce.*—The earliest of all cherries, ripening about the last of May or first of June; it is dwarf in habit, and makes a pretty bush on the mahaleb stock; the foliage is small, dark, and glossy, and it is quite prolific. French.

42. *Jeffries Duke.*—Medium size, red, tender, sub-acid; branches erect and stiff; makes a beautiful pyramid.—Middle of June.

43. *Late Duke.*—Large, light red, late and excellent; tree makes a nice dwarf or pyramid.—End of July.

44. *May Duke.*—An old, well known, excellent variety, large, dark red, juicy, sub-acid, rich; tree hardy, vigorous, and fruitful; ripens a long time in succession; fine for dwarfs and pyramids.—Middle of June, for several weeks.

45. *Morello* (English).—Large, dark red, nearly black, tender, juicy, sub-acid, rich; tree small and slender; makes a fine bush on the mahaleb; if trained on a north wall, it may be in use through all the month of August.

46. *Plumstone Morello.*—Large, dark red, rich and fine; the best of all the morellos; tree a slender, slow grower; makes a nice bush on the mahaleb.—July and August.

47. *Reine Hortense, Monstreuse de Bavay.*—A new French cherry of great excellence; large, bright red, tender, juicy, nearly sweet, and delicious; tree vigorous, and bears well; makes a beautiful pyramid.

NEW AND RARE CHERRIES RECENTLY BROUGHT TO NOTICE.

48. *Bigarreau Monstreuse de Mezel.*—A very large, fine variety, recently introduced from France, but not fully equal to the character given it in the French jour-

nals. It is quite as large as *Tradescant's Black*, and somewhat similar in form; of a dark red, approaching a mahogany color when ripe; very firm; tree of a vigorous habit, similar to the Elton.

49. *Belle d'Orleans.*—A beautiful medium sized pale cherry, ripening immediately after Bauman's May and Early Purple; from France.

50. *Champagne.*—A new variety, raised by Mr. Charles Downing, of Newburgh; described in "Hort.," vol. v., as being very hardy, a great bearer, fruit medium size, brick red, "with a lively rich flavor, a mingling of sugar and acid;" ripe twentieth of June, and hangs long on the tree.

51. *Downing's Red Cheek.*—This is also described in the "Hort.," as "far handsomer, as well as more tender and sweet, than the Bigarreau or Graffion, which it somewhat resembles," and precedes a few days in ripening.

52. *Great Bigarreau.*—This name has been given by Mr. Downing "temporarily, until its real name be found," to a very large, fine cherry, recently brought into notice by Mr. L. M. Ferris, of Orange county, New York, who found it among imported fruit trees growing upon an estate of which he has recently come into possession. It is described as larger than the Black Tartarian, and fully equal in quality, and ripening a few days later; described in "Hort.," in January, 1851.

53. *New Large Black Bigarreau.*—Described in "Hovey's Magazine," December, 1850, as brought from the south of France fifteen or twenty years ago, by a gentleman of Charlestown, Mass. No doubt, identical with "the Great Bigarreau" of Mr. Downing; and as it has been known for many years by this name, it will, of course, take the preference if they prove identical.

54. *Roberts' Red Heart.*—A heart variety, raised in

Salem, Mass., medium size, pale amber, mottled with red, juicy, and sweet; a great bearer; ripe last of June.

55. *Vail's August Duke.*—This is described as being one third larger than the *May Duke*, and ripening at Troy about the eighth or tenth of August; of a bright red color and flavor like the May Duke; originated by Henry Vail, Esq., of Troy; described in "Hort.," vol. iv.

SMALL SELECT LISTS.

For the Garden.—Nos. 11, 13, 4, 44, 12, 9, 34, 35, and 45.

For the Market Orchard.—Nos. 5, 19, 30, 24, 32, 4, and 13.

For Small Hardy Trees.—Nos. 34, 35, 36, 41, 45, 46, 47, and 39.

SECTION 6.—SELECT NECTARINES.

The nectarine tree differs in nothing from a peach, and the fruit only in being smooth skinned. It is peculiarly liable to be destroyed by the curculio, so that it is not advisable to plant it in small gardens.

1. *Boston.*—Large, bright yellow, with a red cheek; flesh yellow, sweet and pleasant flavor, freestone.—First of September.

2. *Downton.*—Large, greenish white, with a dark red cheek; flesh greenish white, rich and high flavored; one of the best. Free.

3. *Early Violet, Violette Hative.*—Medium size, yellowish green, with a purple cheek; flesh pale green, melting, rich and high flavored. Free.—Last of August.

4. *Elruge.*—Medium size, greenish yellow, with a dark red cheek; flesh greenish white, juicy, and high flavored; excellent.—Beginning of September. Free.

5. *Early Newington.*—Large, pale green, red in the sun; flesh pale, red at the stone, juicy, and rich; adheres to the stone. Cling.

6. *Hunt's Tawny.*—Medium size; yellow, with a red cheek; flesh yellow, rich, and juicy—Beginning of August. Free.

7. *Hardwick Seedling.*—Large, pale green, with a violet red cheek; flesh pale green, juicy, melting, and rich. —End of August. Free.

. Nos. 2, 3, and 4, were recommended for general cultivation by the Pomological Congress at New York in 1849.

The *Great Stanwick Nectarine,* of which so much has been said in England, will soon be introduced here, and will be well worthy the attention of those who can give it a wall or a place under glass.

SECTION 7.—SELECT PEACHES.

CLASS I.—FREESTONES.

Fl. s. DENOTES SMALL FLOWERS; gl. GLANDS; glob. GLOBOSE; AND ren. RENIFORM.

1. *Alberge Yellow* (Barnard's, Yellow Rare-Ripe, etc.). —Large, deep yellow, with a dull red cheek, flesh yellow, juicy, and rich; tree vigorous, hardy, and productive.— Beginning of September. Fls. small, globose glands.

2. *Bergen's Yellow.*—Very large, orange, red in the sun; flesh yellow, juicy and fine flavored; tree productive. This is considered one of the best of yellow peaches. —Middle of September. Glands ren. fl. small.

3. *Brevoort,* or *Brevoort's Morris.*—Large, dull white, with a red cheek; flesh pale, sweet, and fine flavored; a good and regular bearer.—Beginning of September. Fls. small, glands ren.

4. *Cole's Early Red.*—Medium size, mostly clouded and mottled with red; flesh pale, juicy, rich, and delicious; tree vigorous, and an abundant bearer.—Middle of August. Glands globose, flowers small.

5. *Cooledge's Favorite.*—A most beautiful and excellent peach; skin white, delicately mottled with red; flesh pale, juicy, and rich; tree vigorous and productive.—Middle to end of August. Flowers small, globose glands.

6. *Crawford's Early.*—A magnificent, large, yellow peach, of good quality; tree exceedingly vigorous and prolific; its size, beauty, and productiveness, make it one of the most popular orchard varieties.—Beginning of September. Glands globose, flowers small.

7. *Crawford's Late Melocoton* (Crawford's Superb).—Really a superb yellow peach, very large, productive and good, ripening about the close of the peach season.—Last of September. Glands globose, fl. small.

8. *Druid Hill.*—Large, roundish, greenish white, clouded with red next the sun; flesh greenish white, juicy, and rich; very productive.—Middle of September. Originated at Baltimore. Fl. s. gl. glob.

9. *Early Newington Free.*—Large, whitish, with a red cheek; flesh pale, red at the stone, rich and vinous flavor.—End of August. Fl. s. gl. ren.

10. *Early Anne* (Green Nutmeg).—Small, greenish white, with a red cheek; flesh pale, sweet, and good.—End of July. Flowers large, no glands, unthrifty, and liable to mildew; only recommended for its earliness.

11. *Early York* (Early Purple, Serrate Early York, etc.).—Medium size; on young thrifty trees large, greenish white, covered in the sun with dull purplish red; flesh juicy, rich and excellent; tree a fair grower and very prolific; one of the best early orchard varieties.—Middle of August. Leaves serrate, flowers large.

12. *Early Tillotson.*—An excellent variety, ripening

with the preceding, about the same size, and of excellent flavor; the tree is sometimes considerably affected with mildew, and in particular cases the fruit also; it should have warm, light soil, and open exposure. Serrate, fls. small.

13. *George the Fourth.*—Large, white, with a red cheek; flesh pale, juicy, and rich; tree vigorous, and bears moderate crops, of the highest quality.—End of August. Gl. glob., fls. small.

14. *Grosse Mignonne.*—Large, dull white, with a red cheek; flesh pale, juicy, with a rich, vinous flavor; a free grower and good bearer. In England it is called " the best peach in cultivation."—End of August. Flowers large, globose glands.

15. *Haine's Early.*—Large, white, with a red cheek; flesh pale, juicy, and delicious; tree hardy and very productive; one of the best varieties.—Middle of August. Fls. small, glob. gl.

16. *Jacques' Rare-Ripe.*—A superb yellow peach, full as large and as good as Crawford's early, and ripening a week or ten days later; origin, Massachusetts. Glands ren. fl. s.

17. *Large Early York.*—A large and beautiful variety, white, with a red cheek; flesh juicy and delicious; tree very vigorous and productive; one of the very best.—End of August. Gl. glob. fl. s.

18. *Late Red Rare-Ripe.*—Large, roundish oval, greyish white, marbled with red in the sun; flesh pale, rich and fine.—Beginning of September. Fl. s. gl. glob.

19. *Late Admirable.*—Large, roundish, oval, yellowish green, with a red cheek; flesh pale, fine flavored.—End of September. Fl. s. gl. glob.

20. *La Grange.*—Large, greenish white, slightly reddened in the sun, flesh pale, juicy, sweet, and rich. Its lateness and color make it a desirable variety for pre-

serving. It should have the warmest soil and situation north of New York, or it will not ripen well.—Last of September or beginning of October; fl. small, glands ren.

21. *Morris' White.*—Medium size, dull creamy white, tinged with red in the sun, flesh white to the stone, juicy and delicious; tree a good bearer; highly prized for preserving on account of the entire absence of red in the flesh.—Middle of September; gl. ren., fl. small.

22. *Morris' Red Rare Ripe.*—Large, roundish, greenish white, with a red cheek, flesh pale, light red at the stone, juicy and rich; trees very productive; fl. small, glands glob.; similar to George IV.

23. *Old Mixon Freestone.*—Large, greenish white and red, flesh pale, juicy, and rich; tree hardy and exceedingly productive; a standard orchard variety.—Middle of September for the north.

24. *Red Rare Ripe* (Ey. Red Rare Ripe).—A fine old sort, whitish, with a dark red cheek; flesh pale, rich, and high flavored.—End of August. Slightly subject to mildew; fl. small; frequently comfounded with the following:

25. *Royal Kensington.*—Very similar to, if not identical with the Grosse Mignonne; several varieties of white fleshed peaches are cultivated about Rochester as the "Kensington."

26. *Royal George.*—Medium to large size, white, with a deep red cheek, flesh white, deep red at the stone, juicy, melting and rich; tree productive.—End of August; fl. small.

27. *Red Check Melocoton.*—A famous, old, well known, and popular variety; large, oval, yellow, with a red cheek; flesh yellow, juicy, rich and vinous; tree very hardy and prolific; valuable for the orchard.—Middle to end of September. Glands glob., fl. small.

28. *Snow Peach.*—A beautiful fruit, medium size, skin

and flesh clear, creamy white throughout; tree hardy and productive, and shoots greenish, very distinct, and one of the most desirable of white peaches for preserving.—Beginning to middle of September; fl. small, white.

29. *Scott's Nonpareil.*—A new, very large and fine yellow peach, from New Jersey, highly esteemed as a valuable market variety.—Middle to end of September; fl. small, glands glob.

30. *Van Zandt's Superb.*—A beautiful smooth fruit, large size, whitish, with a red cheek; flesh pale, juicy, sweet, and good.—First of September; fl. small, glands glob.; origin, Long Island.

31. *Ward's Late Free.*—Large, yellowish white, with a red cheek; flesh pale, juicy, and good; a standard profitable late sort among the Delaware orchardists; will probably not ripen north of New York.

32. *Weld's Freestone.*—A very large, roundish oval, late peach; greenish white, streaked and marbled with red next the sun; flesh pale, pale, juicy, and good; never fails to give an abundant crop at Rochester. Beginning to middle of October; fl. small, glands ren.; succeeds well in Massachusetts.

33. *White Imperial.*—Medium to large size, pale, yellowish white, faintly marked with red; flesh pale, juicy, sweet, and good; tree vigorous; fl. small, gl. glob.

CLASS II.—CLINGSTONES.

34. *Heath Cling.*—A magnificent late peach, cream colored, with a light blush next the sun; flesh greenish white, tender, juicy, and of the highest flavor; fl. small, glands ren.; tree very productive.—Ripe in October; and has the rare property of keeping well for several weeks after being gathered; should be grown on a trellis or wall north of New York to bring it to perfection.

35. *Large White Cling.*—Large, greenish white, lightly reddened in the sun, juicy, sweet, and rich; tree very hardy and productive; highly esteemed for preserving on account of its light color; fl. small, glands glob.

36. *Lemon Cling.*—A very large and beautiful lemon-shaped variety, light yellow, reddened in the sun; flesh yellow, rich, and vinous; excellent for preserving; tree hardy and productive.—End of September. Glands ren., fl. small.

37. *Old Mixon Clingstone.*—Large, round, whitish, with a red cheek; flesh pale, sweet, and rich flavored; fl. small, glands glob.—Beginning of September.

38. *Old Newington Cling.*—Large, yellowish white, with a red cheek; flesh pale, red at the stone, rich, juicy, and good.—Middle of September; fl. large; no glands.

Select lists of Peaches.—Our most profitable orchard varieties in Western New York are, Nos. 1, 4, 6, 11, 15 or 17, 23, 27, and No. 6, the most valuable single variety, on account of its great size and beauty, and the vigor and productiveness of the tree.

Select Garden Varieties.—Nos. 2, 4, 5, 11, 13, 21, 23, and 28.

Robert Manning selects for New England, out of seventy varieties that he has tested, Nos. 11, 13, 22, 6, 2, 19, 37, 7, besides Nivette and Walter's Early. These ten he considers "unimpeachable," and No. 6 he considers combines, in the greatest degree, all desirable quality. With these he recommends Nos. 5, 4, 14, 16, 21, 16, 27, 36, and 32, with several others we have not thought it necessary to describe. He ranks them in regard to relative merit as the numbers are placed.

SECTION 7.—SELECT PLUMS.

1. *Autumn Gage, or Roe's Autumn Gage.*—Medium

size, oval, pale yellow, sweet, juicy and good; parts from the stone; tree a slow grower, but very productive.—Middle to end of September.

2. *Bingham.*—Large and handsome, oval, deep yellow, with a few red spots; juicy and rich; parts from the stone; tree very productive.—Beginning of September.

3. *Bleeker's Gage.*—Above medium size, roundish oval, yellowish; flesh yellow, juicy, and rich; parts from the stone; tree a fair grower and productive.—Last of August.

4. *Cherry, or Early Scarlet.*—Rather small, round, red, very pretty, juicy, soft, sub-acid, adheres to the stone.—Last of July. Makes a very pretty dwarf bush.

5. *Coe's Golden Drop.*—Large and handsome, oval, light yellow, flesh firm, rich and sweet; adheres to the stone; tree a fair grower and very productive, but does not bear so young as many others; valuable not only on account of its large size and fine appearance, but its lateness and hanging long on the tree.—Last of September.

6. *Columbia.*—Large and handsome, roundish, purple, flesh yellow, juicy and rich; parts from the stone; tree vigorous and very productive.—September.

7. *Cruger's Scarlet.*—Medium size, roundish, reddish lilac; juicy, but not rich; an extraordinary bearer; always requires thinning; particularly valuable in light soils; profitable.—September.

8. *Drap d'Or.*—A fine golden yellow plum, somewhat resembling the old green gage; very good; vigorous shoots, a little downy.—Early.

9. *Drap d'Or d'Esperin.*—A new Belgian variety, resembling the Washington, and probably no better. The first trees were sold at Ghent in 1848 at $10 each.

10. *Diamond.*—One of the largest and most productive of purple plums, but coarse; only for cooking.—September.

11. *Dennison's Red.*—Large, round oval, light red, flesh juicy and rich, parts from the stone.—End of August.

12. *Dennison's Superb.*—Pretty large, beautiful, round, yellowish green, with purple dots; flesh juicy, rich, and parts from the stone; tree vigorous and productive.—End of August.

13. *Duane's Purple.*—Very large and handsome, oval, reddish purple; flesh juicy and sweet, adheres to the stone; tree a good grower and very productive.—Beginning of September.

14. *Emerald Drop.*—Medium size, oval, yellowish green; flesh juicy and good, adheres slightly to the stone; a good grower and profuse bearer.

15. *Fellenberg.*—A fine late plum, oval, purple; flesh juicy and delicious, parts from the stone; fine for drying; tree very productive.—September.

16. *Frost Gage.*—Rather small, round, purple; an immense bearer; very late; profitable for market.—October.

17. *Green Gage.*—Small, but of the highest excellence; tree a slow grower.—Middle of August.

18. *German Prune* (Quetsche).—Large, long oval, dark purple, blue, free, fine for drying, and good to eat; grows spontaneously in Germany.—September.

19. *Gen. Hand.*—One of the largest American varieties, introduced by Messrs. Sinclair & Corse, Baltimore. It is of a golden yellow color, sweet but not high flavored.—First of September. Will be valuable for the market, as it is very productive, besides being so attractive in size and beauty.

20. *Guthrie's Apricot.*—Medium size, yellow, has the flavor of the Apricot; of Scotch origin.—End September.

21. *Huling's Superb.*—Large and handsome, round, yellowish green; flesh juicy, rich and fine flavored, parts freely from the stone; tree grows well and is very productive.—Middle of August.

22. *Ickworth Imperatrice.*—An English late variety, purple, flesh juicy, sweet and rich; may be kept into winter.

23. *Imperial Gage.*—Large, oval, greenish; flesh juicy, rich and delicious, parts from the stone; one of the best growers, most productive, and best of plums; profitable for market.—Middle of August.

24. *Ives' Seedling.*—Raised by Mr. J. M. Ives, of Salem, Mass.; large, roundish, oblong, yellow, mottled with red, melting and rich; freestone.—First of September. Tree a strong, rapid grower.

25. *Jaune Hative.*—A nice little yellow plum, ripening last of July; earliness is its chief quality.

26. *Jefferson.*—A new American variety, of the highest reputation; yellow, with a red cheek; flesh orange-colored, juicy and rich, parts from the stone; an excellent variety, but we have never seen it superior to the Imperial Gage.—End of August.

27. *Kirks* (from England).—A large, fine, violet fruit, rich and sugary; freestone.—September. Shoots stout and smooth, like those of the red mag. bon.

28. *Lucomb's Nonsuch* (English).—A large, roundish, greenish plum, nearly as large and as good as the Washington.

29. *Lawrence's Favorite.*—Large, roundish, yellowish green; flesh juicy, melting, and rich, parts from the stone; tree vigorous and very productive.—Middle and end of August.

30. *Lombard.*—Medium size, oval, violet red; flesh yellow, juicy, and pleasant; a great bearer, and said to be peculiarly well adapted to light soils.—End of August. Profitable for market.

31. *Long Scarlet, or Scarlet Gage.*—Medium size, oblong, bright red; flesh juicy, sweet when fully ripe, adheres to the stone; tree a good grower, and a most abundant bearer.—End of August.

32. *Magnum Bonum, Yellow.*—A very large and beautiful egg-shaped yellow plum; a little coarse, but excellent for cooking; tree vigorous and very productive. —End of August. Profitable.

33. *Magnum Bonum, Red.*—Large and beautiful, egg shaped, violet red; of second quality, valuable for cooking; tree vigorous and productive.—End of August. Profitable for market.

34. *Mamelonne.*—A curious looking, distinct fruit; round, with a neck like a pear, greenish, similar in quality and season to the green gage; tree vigorous and productive; new from France.

35. *Mirabelle.*—A small, round, yellow plum, very prolific and fine for preserving.—August and September.

36. *Mirabelle d'Octobre.*—A late variety recently received from France; very hardy and prolific.

37. *Orange.*—One of the largest varieties, oval, yellow, rather coarse; tree vigorous and very productive.—First of October. Profitable for market.

38. *Orleans Early.*—Medium size, round, purple; flesh sweet and good; tree a great bearer.—Middle of August.

39. *Orleans Smith's.*—A very large and excellent variety, oval, reddish purple, with a thick coat of bloom; flesh yellow, firm, juicy, and rich; tree vigorous and very productive.—Last of August. Profitable for market.

40. *Peach.*—A very large and beautiful plum, roundish, dull red; flesh a little coarse; tree very productive.— End of August.

41. *Prune d'Agen, or Robe de Sergent.*—A new French variety, first quality for drying; tree very prolific; medium size, purple, sweet, and good.—September.

42. *Purple Favorite.*—Medium size, brownish purple; flesh juicy, melting, and sweet; one of the very best of plums; tree a slow grower.—Beginning of September.

43. *Purple Gage* (Reine Claude Violette).—Medium

size, roundish, violet, with a blue bloom, rich, sugary, and fine; freestone; hangs long on the tree, and shrivels in ripening: shoots smooth.—September and October.

44. *Red Diaper* (Diaprée Rouge, French).—One of the finest of all plums, brownish red, dark in the sun, freestone. End of August. Hangs long on the tree. This is called Mimms in England, and is different from the Red Diaper of some.

45. *Reine Claude de Bavay* (Esperin).—The best new foreign variety, as large as the Washington, and in flavor equal to the green gage; roundish, oval, greenish, marked with red in the sun; tree vigorous and remarkably productive.—Middle of September. Hangs long on the tree.

46. *Schenectady Catharine.*—Rather below medium size, purple; flesh melting, sweet and excellent.—Middle of August. It is said to reproduce itself from seed without variation.

47. *St. Martin's Quetsche* (German).—Medium size, oval, pale yellow, juicy and rich.—September. Hangs long on the tree; bears the most abundant crops; fine for drying; very profitable.

48. *Thomas* (of Boston).—Large, roundish, oblong, amber colored, juicy and good.—September. Shoots stout, a little downy, a great bearer, and very handsome.

49. *Washington.*—A magnificent, large plum; roundish, green, usually marked with red; juicy, sweet and good; tree vigorous and exceedingly productive; one of the very best.—End of August.

50. *Winter or Late Damson.*—A small, dark purple variety, esteemed for preserving.—October.

51. *Yellow Gage.*—Large, yellow, oval; flesh yellow, juicy, and rich; tree remarkably vigorous and productive; an excellent and profitable variety.—Middle of August.

SMALL SELECT LISTS OF PLUMS.

For the Garden.—Nos. 17, 23, 26, 29, 39, 42, 43, 45, 5.
For Market.—Nos. 16, 23, 32, 33, 39, 49.
For Drying.—Nos. 15, 18, 41, 47.

THIRD DIVISION.—BERRIES.

CURRANTS, GOOSEBERRIES, RASPBERRIES STRAWBERRIES, BERBERRIES, BLACKBERRIES, MULBERRIES, GRAPES, AND FIGS.

SECTION 8.—SELECT CURRANTS.

The currant is a most useful fruit, indeed indispensable to every garden, large or small; it fills a space of a couple of weeks after the strawberries, raspberries, and cherries, and before the apricots, early apples, and pears; and besides this, it possesses such a remarkable combination of sweet and acid, as fits it for an almost endless variety of useful and agreeable preparations, both in the green and ripe state.

The white varieties are mildest flavored, and, therefore, better for using in a raw state when ripe. The red are preferable for jellies, etc., on account of their beautiful color.

1. *Black English*, or common black, well known.

2. *Black Naples.*—The largest and best black currant; bears profusely; valuable for jam and jellies; bunches short, milder flavored, and later than the preceding.

3. *Cherry.*—Largest of all currants, exceeding an inch in circumference, bunches short, color dark red, ripens same time as Red Dutch, shoots stout, short jointed and erect, foliage thick, dark green, slightly folded, and bluntly and coarsely serrated.

4. *Red Dutch.*—A well known variety, bunches three inches long or more; fine.

5. *Red Knight's Sweet.*—Similar to the preceding, but of rather a milder acid.

6. *Victoria or Houghton Castle.*—Very large, bright red, bunches five or six inches long; hangs on the bushes after others are gone; distinguished at once by its remarkably long bunches, and bright red color, and by the foliage, which is quite distinct, dark green, coarsely and bluntly serrated, quite flat, and frequently reflexed or turned backwards at the edges; the shoots are not so stout and erect as those of the cherry.

7. *White Dutch.*—Yellowish white, transparent, milder than the red, and better for using raw; excellent.

8. *White Grape.*—Larger every way than the preceding; the largest white currant; growth rather spreading, foliage thicker, deeper green, and more reflexed.

9. *Missouri Yellow Flowering.*—Fine yellow, fragrant flowers, and sweet fruit of a violet blue.

10. *Missouri Large Fruited.*—Large, blue, sweet fruit, very pleasant.

The two last are seldom cultivated for the fruit.

11. *Long Bunched Red Dutch* (Grosse Rouge de Holland).—This is a variety we received lately from France, and it promises to be valuable. There are several sorts under cultivation, more or less unworthy of notice, unless to those who are making large collections.

Section 9.—Gooseberries.

The following, from the large English sorts, have all proved excellent.

1. *Red.*—Albion, Crownbob, Echo, Houghton's Boggart, Ironmonger, Lancashire Red, Prince Regent, Roar

ing Lion, Shakspeare, Sportsman, Top Sawyer, Wineberry, Young's Wonderful.

2. *White.*—Chorister, Fleur de Lis, Leigh's Toper, Queen Caroline, Smiling Beauty, Whitesmith, Wellington's Glory, White Muslin, etc.

3. *Green.*—Berrier's Greenwood, Chipendale's Conquering Hero, Green Mountain, Green Vale, Green Willow, Green Ocean, Independent, Jolly Cutler, Massey's Heart of Oak, Profit.

4. *Yellow.*—Bunker Hill, Capper's Early Sulphur, Golden Drop, Husbandman, etc.

5. *Houghton's Seedling.*—Raised in Massachusetts from the seed of a native variety; it is small and rather indifferefit in flavor, but is not subject to the mildew, and bears most abundantly, small, oval, dull brownish red.

The following sorts were recommended by the Pomological Congress: Houghton's Seedling, Whitesmith, Crownbob, Red Champagne, Warrington, Laurel, Ironmonger, Early Sulphur, Green Gage, Green Walnut.

SECTION 10.—SELECT RASPBERRIES.

1. *Antwerp, Red.*—This is an excellent variety, and very popular in market; three quarters of an acre of land on the Hudson, planted with it, have yielded $330; and three acres in the same locality, $1,500 in one season. The berry is large, conical, dark red, rich and juicy; canes have a few small, purple spines.

2. *Antwerp, Yellow or White.*—Fruit large, pale yellow, sweet and rich; a beautiful and excellent fruit, but not so firm and so well adapted to marketing as the preceding; canes thickly covered with greenish spines.

3. *Fastolff.*—Fruit larger and rounder than the Red Antwerp, but rather softer; of a purplish red, canes more spiny; very hardy and productive.

4. *Franconia.*—Fruit very large, of a purplish red, rather darker than the Red Antwerp or Fastolff; canes very strong, with a few short purple spines, and thicker, firmer and smoother, or less crimped or wrinkled leaves than any of the others.

5. *Knevett's Giant.*—This is an English variety of the Red Antwerp, from which it differs only in being somewhat hardier.

The " *Col. Wilder*" and " *Cushing*" are two seedling varieties produced by Dr. Brinkle, of Philadelphia, that give promise of superiority, the first especially, which is described as a beautiful " cream-colored" fruit and very hardy.

The American Red, White and Black are well known.

6. *Large Fruited Monthly* (New).—Large red, bears in favorable weather from August to November; canes long, rather slender, purplish in the sun, and pretty thickly covered with dark purple spines; quite different from the " Ohio Everbearing," a worthless variety.

We saw fruit on this in January, in Mr. Rivers' nursery, in 1848. If the autumn be dry, the plant should be watered occasionally; and to ensure a good autumn crop, the canes should be pruned in spring to within a foot of the ground.

SECTION 11.—SELECT STRAWBERRIES.

Those strongly pistillate or deficient in stamens are marked (p).

1. *Alpine, Red Monthly.*—Small, high-flavored variety, and highly valuable in all collections, on account of bearing a long time.

2. *Alpine, White Monthly.*—As above, except color.

3. *Alpine, Red Bush.*—These have no runners, like

other varieties, and are well adapted to edging walks in the kitchen garden. They are small, but of delicious flavor, and continue bearing till autumn.

4. *Alpine, White Bush.*—Same as above, but in color.

5. *British Queen* (Myatt's).—The most magnificent in appearance of all strawberries, often measuring six or seven inches in circumference; but it is a shy bearer and rather tender; plant very luxuriant.

6. *Bishop's Orange.*—Rather large, light orange scarlet, productive and fine flavored (p).

7. *Boston Pine.*—Large light red, good, hardy and exceedingly productive; with plenty of room and good culture the yield is very great.

8. *Burr's Seedling.*—A very prolific medium sized variety; hardy and of fair quality; from Ohio.

9. *Burr's New Pine.*—Large, light orange, scarlet; of the highest and most delicious flavor uniformly; plant hardy and productive; one of the very best sorts (p) Ohio.

10. *Burr's Rival Hudson.*—Medium size, dark red, rather acid; valuable for marketing and preserving; hardy and productive (p). Ohio.

11. *Burr's Columbus.*—Large, hardy and productive; flavor medium (p). Ohio.

12. *Burr's Scarlet Melting.*—A very pretty light scarlet fruit, and a most profuse bearer, but very tender; not fit for marketing.

13. *Black Prince.*—A large and beautiful fruit, of a dark blackish crimson color; variable in quality; some seasons first rate, others insipid; hardy and productive (p).

14. *Climax Scarlet.*—Medium size, conical, slightly necked, light scarlet, rather acid; bears immense crops (p).

15. *Duke of Kent.*—Small, with a long neck; very prolific and valuable for its earliness in a large collection.

16. *Genesee.*—Large, roundish, dark crimson, good;

plant very luxuriant; fruit stalks very stout, supporting the fruit well; most profuse bearer, rather late.

17. *Hudson.*—Medium size, scarlet, firm, acid; very productive, and esteemed for marketing; grown much around Cincinnati (p).

18. *Hovey's Seedling.*—A well-known magnificent berry; plant hardy and luxuriant; bears large crops in some places and seasons (p).

19. *Jenny's Seedling*—Very large, roundish, dark scarlet, flavor medium, plant vigorous, and a moderate bearer (p).

20. *Large Early Scarlet.*—An excellent standard sort, light scarlet, rather acid; bears uniformly great crops; early.

21. *Monroe Scarlet.*—Large, roundish, light scarlet, good; very prolific; over 100 perfect berries have been gathered at once from a single plant (p).

22. *Orange Prolific.*—Large, orange scarlet; rather acid, but a great bearer and quite late.

23. *Princess Alice Maud.*—A very large and handsome English variety; very productive, but of indifferent flavor.

24. *Prolific Hautbois.*—A large, purplish, conical fruit, with a peculiar musky flavor, very productive; plant grows tall and luxuriant, with peculiar crimped foliage, and has very large, showy blossoms; a distinct species; late.

25. *Swainstone's Seedling.*—A very large and beautiful fruit, of the most delicious flavor; color light shining scarlet, ripens gradually; a poor bearer; English.

26. *Scotch Pine Apple, or Crimson Cone.*—One of the most beautiful varieties in appearance; medium size, uniform, regularly conical, rich dark crimson; seeds deeply imbedded, giving the surface a rasp-like appearance; rather acid but good, and very productive.

Nos. 7, 9, 18 and 20 are the best for general cultivation; for a larger collection, Nos. 5, 17 and 25, besides the alpines, may be added; Nos. 14, 16, 21, 22 are new seedlings produced here that promise well.

The wood and alpines should be renewed from seed frequently.

Section 12.—Berberries.

EPIN.-VINETTE OF THE FRENCH.

Common Red.—This is everywhere well known; grown not only for the fruit, which is used for preserves, jellies and pickles, but for ornament. The bright scarlet oval fruit is borne in rich clusters, and hang on till late in the autumn.

Sweet-Fruited (Berberis dulcis).—The fruit of this is much less acid than the common. The plant is not so vigorous.

Besides these, there are several species and varieties cultivated chiefly for ornament: *The White-fruited*, *The Violet-fruited*, *The Variegated-leaved*, foliage marked with yellow; *The Purple-leaved*, the most unique and ornamental of all, with beautiful violet-purple foliage.

They are all easily propagated by layers or suckers, and the rare sorts by grafting.

Section 13.—Blackberries.

The Improved High Bush.—This Blackberry is beginning to receive considerable attention. The Massachusetts Horticultural Society has offered large premiums to encourage its culture, and the result already has been great improvement. Capt. Lovett, of Beverly, has present specimens an inch and a half long. It bids fair to become a valuable and popular fruit. The berry is long, egg-shaped, shining black, juicy, and rich, the plant erect, blossoms white, ripens at a most timely season, after the Raspberry.

Section 14.—Mulberries.

Black.—This is a native of Persia, and is really the only one valuable for its fruit. The berry is an inch and a half long, and nearly an inch in diameter, black, succulent, sugary and rich. The tree is highly ornamental, very erect, with a large spreading head. The leaves appear late in spring, are large, heart-shaped, sometimes lobed, deep green, and form a dense shade.

Section 15.—Grapes.

SELECT HARDY GRAPES.

1. *Catawba.*—This is the best flavored of all native grapes that ripen as far north as lat. 43 deg., and is considered the best yet discovered for making wine. Bunches large; berries large, red, becoming a coppery color when ripe; juicy, sweet, and musky; hardy, and very productive.

2. *Clinton.*—A very hardy, native variety, resembling in foliage the common Fox Grape. Bunches small and very compact; berries rather small, black, juicy, inferior in flavor to the preceding. It ripens here two or three weeks before the Isabella or Catawba, and this is its chief value; very productive.

3. *Diana.*—This is a variety that originated near Boston, similar to the Catawba; not quite so large, but earlier and better adapted to the north.

4. *Isabella.*—This is the most popular variety. It ripens well in almost every part of the country, and bears immense crops under the most ordinary management.

Bunches long and large; berries large, oval, black, juicy, sweet, slightly musky.

SELECT FOREIGN GRAPES.

1. *Black Cluster.*—Small, roundish oval, black, sweet and good; bunches small, very compact; one of the hardiest and best for open air culture; early.

2. *Black Prince.*—Large, oval, black; bunches long, rather open; sweet and fine; a profuse bearer.

3. *Black Frontignan.*—Berries medium size, round, black, bunches long; flavor rich and musky; prolific.

4. *Black Hamburg.*—A fine grape, and a general favorite for the vinery; bunches are large, very much shouldered—that is, branched; berries large, deep black, sweet and rich.

5. *Chasselas de Fontainbleau.*—This is esteemed the finest table grape in France, and succeeds admirably here in vineries, and occasionally in the open air. Bunches large, somewhat shouldered; berries large, round, greenish white, becoming slightly colored or reddened in the sun; canes stout, of a yellowish color; leaves large and shining; very productive. The Golden Chasselas is very similar to, if not identical with this.

6. *Grizzly Frontignan.*—This is one of the most delicious grapes when grown in the vinery, and very beautiful too. Bunches long, slightly shouldered; berries medium size, round, colored red and violet-purple in the sun; rich, musky flavor.

7. *White Frontignan* (Muscat Blanc of the French).—One of the oldest varieties; bunches pretty large; berries roundish, changing from green to amber as they ripen in the sun; rich and quite musky; later than the preceding.

8. *White Muscat of Alexandria.*—This is a most delicious variety, considered the same as the imported

"Malaga." Bunches large, branched and loose; berries large, oval, white, becoming amber; firm and rich, with a high musky flavor; growth vigorous; leaves shining and deeply lobed.

9. *White Sweetwater.*—This and the Black Cluster are the most common foreign varieties in this country. Bunches of good size, open; berries of medium size, round, green, becoming slightly colored in the sun; sweet and watery; occasionally produced in tolerable perfection in the open air.

The Pomological Congress at New York, in 1849, recommend for culture under glass, Nos. 2, 3, 4, 5, 6, 7, 8.

Section 16.—Figs.

Very little is known here from experience of the particular qualities of the different kinds of figs. Several varieties have been tested, and are grown successfully in the Boston graperies. The varieties most desirable for out-door culture for their hardiness are:

The Brown Turkey.—Large, oblong, pear-shaped; skin dark, brownish purple; flesh red, leaves large.

Black Ischia.—Medium size, roundish, dark violet, nearly black; flesh deep red, sweet and fine. This is one of the most productive varieties.

Violette de Bordeaux (Figue poire de Bordeaux.—Large, long, pear-shaped, brownish red; flesh reddish, medium quality; extensively cultivated about Paris for its productiveness and hardiness.

White Marseilles.—Small, roundish, nearly white; flesh white; not quite so hardy as the preceding sorts, but very productive; one of the most abundant in the Paris markets.

Upwards of forty varieties are described in the London

Horticultural Society's catalogue; fifteen are described in Downing's Fruit and Fruit Trees.

FOURTH DIVISION.—ALMONDS, CHESTNUTS, FILBERTS, AND WALNUTS.

SECTION 17.—ALMONDS.

1. *Sweet Hard Shell.*—This is a hardy and productive variety, succeeding well in the climate of Western New York, and still farther north. Nut very large, with a hard shell and a large sweet kernel; ripe here about the first of October.

The tree is very vigorous, has smooth glaucous leaves, and when in bloom in the spring, is more brilliant and showy than any other fruit tree.

2. *Soft Sweet Shell, Ladies' Thin Shell, etc.*—This is *the* almond of the shops, of which such immense quantities are annually imported from abroad. It and all its sub-varieties, as far as we know, are too tender for our northern climate, unless carefully grown on a wall or trellis, and protected. South of Virginia, we believe, it succeeds well; and so beautiful a tree, and so estimable a fruit, deserve the attention of all fruit growers.

3. *The Bitter Almond.*—This is hardy and productive; nut similar to No. 1 in appearance, but bitter, and only useful in confectionery or medicine. Its chief product is the prussic acid of the druggists.

SECTION 18.—CHESTNUTS.

The American or Common Chestnut is well known as one of our most beautiful forest trees. It is seldom grown as a fruit tree, although the fruit is highly esteemed.

It should have a place in all large collections of standard fruit trees. It reproduces itself from seed.

The Dwarf Chestnut or Chinquapin, is a small tree eight or ten feet high, and very prolific, but the nuts are small. It grows spontaneously in Maryland, Virginia, and southward.

The Spanish Chestnut or Marron.—This is the large, sweet nut, as large as a horse chestnut, imported from abroad. There are many varieties cultivated in France and England, but that designated by the French as " *Marron de Lyon*," is the best. It is propagated by grafting on the common sorts. It is not reproduced truly from seed, but its seedlings produce large and fine fruits. It bears and ripens well as far north as Rochester. It bears the second year from the graft and the fourth from seed.

Section 19.—Filberts.

1. *Cosford.*—This is an improved variety of the English hazel-nut, very prolific, nut large, oblong or oval, shell thin, and kernel fine flavored.

2. *Coburg.*—Large and fine, and a most abundant bearer.

3. *Dwarf Prolific.*—One of the most prolific bearers, nut rather small. We have plants two feet high bearing well, kernel good.

4. *Frizzled.*—Remarkable for its curious frizzled husk, a good bearer, and one of the finest flavored.

5. *Red Skinned.*—One of the old standard sorts of the English growers, distinguished by the bright red or crimson skin of the kernel, medium size, egg-shaped, shell thick, flavor good.

6. *White.*—This is also an old standard sort, the kernel is a yellowish white. Both this and the preceding have long husks.

Section 20.—Walnuts.

The English or *Madeira Nut* (Juglans Regia).—This is a native of Persia. A lofty spreading tree with pinnated leaves like the butternut, and the fruit nearly as large. Great quantities are annually imported, and sold in the fruit shops.

The tree is tender while young, the ends of the young shoots being injured in winter at the north, but as it grows older it becomes hardier. It is produced from seed or by grafting. There are many varieties of it cultivated abroad, few of which have yet been introduced here on account of the little attention given to this class of fruits.

The Dwarf Prolific Walnut (Juglans Præparturiens), is a French variety recently introduced, which will probably become the most desirable for the garden. It bears at the age of three years from the seed, and often at the height of two to three feet. We have now two imported trees, four feet high, that give promise of an abundant crop. The kernel is said to be very good.

Our native sorts, the Black Walnut (Juglans Nigra), the Butternut (Juglans Cinerea), the Hickory Nut (Carya), and its varieties, are all well known trees that deserve much more attention than they receive, considering the value of their timber as well as fruit.

CHAPTER II.

GATHERING, PACKING, TRANSPORTATION AND PRESERVATION OF FRUITS.

THIS is a branch of the general subject of fruit culture and management that requires the most careful attention; for it is quite useless to take pains in producing fine fruits, without taking equal pains in gathering, preserving, and sending them to the table or the market in a sound, sightly, and proper condition. Very few fruit growers seem to appreciate this part of their business. Fruit dealers at home and abroad complain of the careless and slovenly manner in which our fruits are gathered, packed, and presented in the market, and would gladly pay a double price for them in a better condition. The first consideration is—

The period of maturity at which fruits should be gathered.—The stone fruits generally are allowed to reach perfect maturity, or within four or five days of it, on the tree.

In moist, cool seasons particularly, they are benefited by being gathered a few days before maturity, and allowed to ripen in a dry, warm room; they part with the water contained in their juices, which thus become better elaborated and more sugary and high flavored.

Summer Pears, too, on the same principle, require to be gathered, as a general thing, from a week to a fortnight before their maturity. Sweet varieties, and such as are inclined to become *mealy*, are entirely worthless when ripened on the tree, and many very excellent varieties are condemned on this account. Such as these should be gathered the moment the skin begins to change color in least degree.

Summer Apples, too, and especially those inclined to *mealiness*, should be picked early; as soon as the skin begins to change color, otherwise they part with their juices, and become worthless. Ripeness is indicated by the seeds turning dark colored, and by the stem parting readily from the tree when it is lifted upwards.

Winter Apples and Pears should be allowed to remain on the trees as long as vegetation is active, or until frosts are apprehended.

Grapes, Berries, &c., are allowed to attain perfect maturity before being gathered.

Chestnuts, Filberts, &c., are not gathered until they begin to fall from the tree.

Mode of Gathering.—Unless it be a few specimens wanted for immediate use, which may be taken with some of the contrivances mentioned under the head of implements, all fruits should be gathered by *the hand*. The branch to be gathered from should be taken in one hand, and the fruits carefully taken off, one by one, with the other, with their stems attached. (For fruits neither keep so well, nor look so well, without the stems.) They are then laid carefully in single layers in broad shallow baskets, the bottoms of which should be covered with paper or moss, to prevent bruises. Peaches and other soft fruits should be pressed as lightly as possible, for anything like a squeeze is certainly followed by decay in the form of a brown spot, and this is the reason why it is so exceedingly difficult to find a perfectly *sound* and at the same time *ripe* peach in our markets.

When more than one layer of fruit is laid in the same

basket, some soft paper, dry moss, hay, or other material, ought to separate them, for it is difficult to place one layer immediately upon another, and especially if the fruits are approaching maturity, without bruising them more or less. Fruit should only be gathered in dry weather, and in the dry time of the day.

Disposition of the Fruits after gathering.—When they are thus in the baskets, if summer fruits, they are either carried into the fruit room and arranged on shelves or tables in thin layers, or they are carefully transferred one by one into market baskets and carried to market on an easy spring wagon, if not by steamboat or railroad, by which jarring or jolting will be avoided. Treated in this manner, they will be in a marketable condition, and one basket will sell for as much as four, carelessly picked, thrown into baskets, and tumbled out of them into a barrel or wagon-box.

Ripe fruits may be kept in good condition for a considerable period of time, in an ice-house, or in some of the recently-invented fruit preservers, and even in very cool dry cellars. The vessels in which they are deposited, should be perfectly clean, that no unpleasant flavor may be imparted to them. Peaches have been sent to the East Indies, by being properly packed in ice; and it may be that methods of packing and preserving will, before long, be discovered, that will give us access to the markets of other countries, even for our perishable summer fruits. We have seen Seckel pears in a very good state of preservation in January, exhibited in the horticultural society's rooms in Boston. The science of ripening and preserving fruits is but in its infancy, and horticultural societies that have the means will be doing a great public service by offering liberal premiums that will incite to experiment on the subject.

Winter Fruits intended for long keeping are transferred

by hand from the baskets in which they are gathered on the tree, into larger ones in which they can be carried into a dry cool room, where they are laid in heaps, which may be three or four deep, where they may remain for a couple of weeks, during which time they will have parted with considerable moisture and be quite dry. They will then be fit for packing.

Clean, new barrels should be procured, and the fruits should be carefully assorted. For shipping to distant or foreign markets, *the best only* should be selected; all bruised, wormy, knotty specimens being laid aside for home consumption. They are then placed in the barrels, by hand, arranged regularly in layers, so that no spaces will éxist, by which the fruits may shift, roll, or knock against one another. The barrels are then tightly headed up, so that the head presses firmly on the fruits; some people recommend placing a layer of clean moss or soft paper, both on the bottom and top of the barrel; but this is not necessary where the packing and heading are performed carefully. After packing, the barrels must be sent to market in such a manner as never to be jolted or rolled, any more than they would be on men's shoulders, or an easy spring wagon or sled, or by a water conveyance.

On shipboard the barrels should be placed in the coolest and dryest place. It is perfectly idle to gather, pack or ship fruits in any other way than this to foreign markets. American apples are frequently sold in Liverpool at auction for half what they would have sold for in New York, on account of their bad condition. I saw this in 1849, when Newtown pippins were selling at twelve and a half cents a-piece in the fruit shops.

Winter fruits for home consumption should be care fully assorted, keeping the best, the poorest, the sound, the bruised, and the earlier and later ripening varieties all separate; when sound and bruised, early and late, are all thrown together promiscuously, they cannot fail to decay speedily and to lose their flavor; for two or three decaying apples in a heap or barrel will taint the flavor of all, and hasten the decay of those around them. This arrangement into grades and classes is, therefore, absolutely necessary even for the fruits needed for family use; and when they are so arranged, the sound, long keepers are put into clean, new barrels, carefully by hand, and the barrels headed up tightly and placed in a cool dry cellar or fruit room. The bruised ones can be laid in a place by themselves for immediate use. Every barrel, when packed, should be marked.

Winter Pears, as a general thing, require to be brought into a warm temperature one or two weeks before they are wanted for table use. All the baking and stewing, and even many of the table varieties, may be treated exactly like apples.

Packing Pears for distant markets.—The French send away more pears to foreign markets than any other people. Some small importations of their winter sorts have actually been made by some of the New York fruit dealers the present winter, 1850-51. They pack them in small boxes, either round or square, such as a man can lift and carry easily in his hands.

They cover the bottom and sides with very dry moss or soft dry paper, well calculated to absorb moisture. They then wrap each fruit in the dry, soft paper, and lay them in layers, the largest and least mature in the bottom, and fill all the interstices with dry moss or paper. I have seen these boxes opened in London, in the finest condition, after being a month packed. They are so tightly packed that the slightest movement cannot take place among them, and yet no one presses upon another. The dry moss and paper that separate them, absorb any moisture; and if one decays it does not affect others.

Some of the Paris confectioners and restaurant keepers preserve fruits very successfully in barrels, packed in layers, and the interstices filled up with powdered charcoal. The barrels are kept in a dry, cool place, about forty degrees, where they are not subjected to changes of temperature. Apples, pears, grapes, almonds, nuts, and potatoes, are all preserved in this manner.

Fruit Rooms.—A fruit room is a structure set apart exclusively for the preservation of fruit. Its great requisites are, perfect security from moisture or dampness, exclusion from light, and a uniform temperature. If these points are obtained, no matter where, how, or of what material the fruit room be constructed. It may be built of stone, brick, clay, or wood, above or below ground, as circumstances or taste may dictate.

A good, dry, and cool cellar, is as good a place for keeping fruit in as can be provided; but the great objection to cellars used for other purposes is, that currents of air are frequently admitted, and too much light, by which the temperature is changed, decay promoted, or the fruits dried and shrivelled. There are, also, other objects that unavoidably saturate the air more or less with moisture.

Where a fruit room is built on the surface of the ground, it should be on the ice-house principle of double walls and doors, to prevent access of either heat or cold from without. A good cellar or cave, built in a dry, sandy, or gravelly bank, or side hill, will answer every purpose. The walls may be of stone, brick, or timber; the roof should be thick, with a slope sufficient to throw off water freely, and the earth about should also be so graded that water will flow away as fast as it falls. Provision may be made for lighting and ventilating in the roof, and the door or doors should be double.

The interior should be fitted up with shelves and binns, with places for barrels or other articles, in which fruits are packed.

Attention to fruits in the cellar or fruit room.—The decay of fruits is caused either by bruises or by a fungus, or species of mildew, that increases rapidly and attacks all the sound fruits within its reach. It is, therefore, necessary to examine fruits frequently, and remove all that show any symptoms of decay, before they have either affected others or tainted the atmosphere of the room.

CHAPTER III.

DISEASES AND INSECTS.

SECTION 1.—DISEASES.

1. *The Fire Blight of the Pear, Apple, and Quince.*— This is one of the most formidable diseases to which fruit trees are liable. Whether it is caused by the sun, the atmosphere, or an insect, remains in doubt, some cases favoring one opinion, some another. It attacks the trees at different periods of the growing season, from June to September, and generally in the young parts first; the leaves flag, the sap becomes thick and brown, oozing out in globules through the bark, and emitting a very disagreeable odor, and the diseased branch or part turns black, as if it were burned by fire. When the pear tree is attacked it is difficult to save it, the disease spreads so rapidly. In the apple and quince it is less fatal, rarely killing more than a portion of the tree even if left to its own course. The only remedy is, to cut away instantly the blighted parts, into the sound wood, where there is not the slightest trace of the disease, and burn them up immediately.

It is thought by some that young trees growing very rapidly are more subject to it than older trees growing slowly; and that warm sunshine, with a sultry atmosphere after rain, is apt to be followed by much blight. We have always regarded the cases favoring such an opinion as accidental.

2. *Pear leaf Blight.*—This disease has already been alluded to in treating of pear seedlings. It is a sort of rust that appears on the leaves in July or August, first as small brown spots; these spread rapidly over the leaves until they are completely dried up and growth stopped. It appears in a certain spot as a centre, from which it spreads. Whether it be an insect, a fungus, or some atmospherical cause that produces this blight, is unknown. Certain cases favor one or other of these opinions. More minute investigations are wanted on the subject.

To avoid its evil effects as far as possible, the great point is, to get a rapid, vigorous growth, before midsummer, when it usually appears. Seedlings grown in new soils do not appear to be so much affected as in old. Where stocks are affected very early in the season, they become almost worthless, on account of the feebleness produced in both stem and roots by such an untimely and unnatural check. Some *special* applications, such as coal cinders, iron filings, copperas, etc., have been suggested, but no evidence has yet been produced of their efficacy.

3. *The Gum in stone fruits.*—The cherry, plum, apricot, and peach, are all more or less subject to this malady. The cherry is particularly liable to it in the West. It is produced by different causes, such as a wet soil, severe pruning, pruning at an improper time, violet changes of temperature, etc. The gumming of the cherry in the West, is considered by some to be owing in a great measure to the bark not yielding naturally to the growth of the wood, and hence they practise longitudinal incisions on it. The cherry tree has a very powerful bark, and in some cases it may not yield naturally to the expansion or growth of the wood. We have seen about a foot of the trunk of a cherry tree, several inches smaller than the parts both above and below it. The bark was as smooth as glass on it, the first rind being unbroken,

whilst on the large parts this was quite rough. This was a case arising from the obstinacy of the bark, and could only be remedied by longitudinal incisions on the small part.

It is most probable that the extent and severity of this disease in the West is owing to violent changes from a hard frost to a bright sun and rapid thaw, by which the sap becomes deranged, and accumulates in masses. Trees that are branched near the ground, will be less likely to suffer than those with tall bare trunks. Where it has made some progress in any tree, the only remedy is to pare off the diseased bark, clean off all the gum, and let the surface dry up; then apply a plaster of grafting composition, or a solution of gum shellac in alcohol, put on with a brush, as recommended by Mr. Downing.

When the stone fruits are pruned severely in the spring, the sap does not find sufficient vent; it accumulates in masses and bursts the bark. This fact should always be kept in view in pruning, and a sufficient supply of active buds be left to absorb the sap.

4. *The Yellows in the Peach.*—This is supposed to arise from negligent cultivation. It exhibits itself in a yellow, sickly foliage, feeble shoots, and small fruits prematurely ripened. It is said to be contagious. Trees exhibiting these symptoms should instantly be destroyed. To avoid it, care should be taken to propagate from trees in perfect health and vigor.

5. *Mildew on the Peach.*—The young shoots, leaves, and even the fruit of certain varieties, and especially the glandless ones, such as *Early Anne, Early Tillotson*, etc., are attacked by this. The only remedies are, to give the trees a dry, good soil, that will keep them in a vigorous condition, and to syringe freely twice a day when it begins to appear. The gooseberry suffers seriously from the mildew, owing mainly to the heat of our summers. In

Northern New York, in Maine, Vermont, and Lower Canada, the finest large English varieties are brought to greater perfection than in warmer districts, and with good culture almost come up to the English standard. In a cold, damp-bottomed soil at Toronto, almost on a level with Lake Ontario, fine crops are produced with comparatively little difficulty from mildew or rust. This would indicate as a remedy, a cool soil and situation, and mulching the roots to keep them cool. The plants should be renewed every three or four years, and they should be kept vigorous by liberal manuring and good culture.

6. *The Plum Wart or Black Knot.*—The cause of this disease is quite uncertain, but the probability is that it originates in a similar way to the gum, from an imperfect circulation of the sap, induced by violent changes of temperature.* Cutting out the diseased branch clean to the sound wood, the moment the knots begin to appear, is an effectual remedy, and they should all be burnt up. We have saved trees six inches in diameter, that were affected on the trunk so seriously, that one third of its thickness had to be removed to get below the disease. After it was cut out, we applied a plaster of grafting composition, covered it with a cloth, and in two years it was all healed over and sound.

Plum trees are so neglected in the country, that multitudes of them are now standing literally loaded with these warts—not even an inch of any branch free from them—the most disgusting objects in the way of fruit trees that can possibly be imagined.

6. *The Curl of the leaf in the Peach.*—This disease causes the leaves to assume a reddish color, to become

---

* We have observed that cold weather, about the blossoming period, induces the gum in plum trees as well as in the peach, and when it continues long, as in 1849–50 in Western New York, there is an unusual development of it.

thi. k, curled, and deformed, and finally to perish. It is supposed by many to be caused by insects;* but it is really induced by a sudden change of weather.

A number of warm days, that cause the expansion of the young leaves, followed by a cold rainy day, is almost sure to produce it to some extent; and the more severe and protracted the cold, the more extensive and fatal it is. The peach trees in Western New York suffered more from this in 1849–50 than in the ten years previous, owing to a protracted cold time in each season after the young tender leaves had expanded. In both these seasons the check was so severe, as not only to produce this disease in its worst form, but the *gum* also; for the sap not being absorbed by the leaves, became stagnant, sour, and corroded, and burst the bark. Trees in sheltered gardens suffer less than those in exposed orchards. There is no possible way of guarding against this; and the only remedy known to us is, to pick off the diseased leaves the moment the weather changes, that new healthy ones may be produced.

SECTION 2.—THE PRINCIPAL INSECTS INJURIOUS TO FRUIT TREES.

1. *Aphis or Plant-Louse.*—There are several kinds of these. The two most troublesome to fruit trees are the green and black, small soft insects that appear suddenly in immense quantities on the young shoots of the trees, suck their juices, and consequently arrest their growth. The apple, pear and cherry, are especially infested with them. They multiply with wonderful rapidity. It is said that one individual in five generations might be the progenitor of six thousand millions. Were it not that they

---

* Prof. Harris says in his Treatise on Insects, that it is caused by plant-lice puncturing the under sides of the leaves.

are easily destroyed, they would present an obstacle almost insuperable in the propagation and culture of trees.

There are many ways of accomplishing their destrucion. Our plan is to prepare a barrel of tobacco juicè, by steeping stems for several days until the juice is a dark brown, like strong beer; we then mix this with a solution of soft soap or soap suds. A pail is filled with this, and the ends of the shoots where the insects are assembled are brought down and dipped into the liquid. One dip is enough. Such parts as cannot be dipped are sprinkled liberally. It is applied to the heads of large trees by means of a hand or garden syringe. It should be done in the evening. The liquid may be so strong as to injure the foliage, hence it will be well for persons using it the first time to test it on one or two subjects before applying it extensively. This application must be repeated as often as any of the aphides make their appearance. The dry weather of midsummer is generally the time most favorable for their appearance.

2. *The Woolly Aphis or American Blight.*—This is a small insect, covered with a white woolly substance that conceals its body. They infest the apple tree in particular, both roots and branches, living upon the sap of the bark, and producing small warts or granulations on it by the punctures. They are more particularly troublesome on old rough-barked trees, as they lodge in the crevices, and are difficult to reach. The wind carries them from one place to another by the light down in which they are enveloped, and thus they spread quickly from one end of a plantation to the other. Not a moment should be lost in destroying the first one that makes its appearance. Where the bark is rough it should be scraped smooth, if the roots be affected the earth should be removed, and every part washed, and every crevice filled with the following preparation, recommended in Harris's Treatise:

"Two parts of soft soap and eight of water, mixed with lime enough to bring it to the consistency of thick whitewash, to be put on with a brush." A solution of two pounds of potash in seven quarts of water will answer the same purpose. Fresh earth should be put upon the roots.

3. *The Scaly Aphis or Bark Louse.*—This is a dark brown scale insect, that infests the bark of the apple tree. They are of a dark brown color just like the bark, and are not easily seen unless looked for. They attach themselves closely to the bark, and sometimes are so numerous as to form a complete coating. They seldom appear on thrifty growing trees in good soil; but where the soil is damp and cold, and the trees growing feebly, this insect may be looked for. June is the time to destroy them, when they are young. At other times they are hard, and able to resist any ordinary remedy. The same application recommended for the aphis, applied to them with a hard brush, will effect their destruction. Where they have been left for a long time undisturbed, and have pretty well covered the tree, the quickest and best remedy is *to destroy tree and all*, unless it possesses some extraordinary claim for indulgence. Prof. Harris mentions having found a reddish brown bark louse on his grapevine, arranged in rows one behind another in the crevices of the bark.

4. *The Apple Tree Borer* is a very troublesome insect in some sections of the country. In Western New York we have never met with it but in two or three instances, in very old, neglected orchards, that had stood for twenty years in grass. The beetle is striped brown and white, and is about three-fourths of an inch long. It deposits its eggs in June, in the bark of the trees near the ground. Here the larva is hatched, becoming a whitish grub, which saws its way into the tree, perforating it in all

directions, sometimes completely girdling it. The most effectual method to destroy them is, to insert the end of a wire into their burrow, and killing them. The same means are taken to guard against them as against the peach tree grub, viz., placing a mound of ashes around the base of the trunk in the spring, and allowing it to remain until after the season in which the beetles deposit their eggs. It prevents them from reaching the soft bark at the surface of the ground, the place usually selected. It is stated in Downing's Fruit and Fruit Trees, that "the beetles may be destroyed in June by building small fires of shavings in different parts of the orchard."

5. *The Apple Worm.*—The apple moth deposits its eggs in the eye or calyx of the young fruit; the grub is there hatched, and eats its way into the fruit, leaving behind it a brownish powder. Sometimes the apples drop before they are half grown, and occasionally remain until they acquire a premature ripeness. Early apples are more affected, generally, than late ones, probably because in a more forward state when the eggs are deposited.

When the fruit falls the grub immediately leaves, prepares itself a place in some crevice of the bark of the tree, and spins a thin paper-like cocoon, in which it spends the winter, to come out the following spring and reproduce itself. There are but two ways of destroying them; one is, at pruning time in March, to search carefully for the cocoons and destroy them, and the other is to pick up promptly all fallen wormy fruits and destroy them. These two means, industriously followed, will greatly diminish the amount of wormy fruit, the increase of which is exciting alarm.

6. *The Canker Worm.*—This insect is confined chiefly to New England; we have never seen it in New York.

They generally emerge from the ground in March. According to Professor Harris, some rise during the late autumn and winter months. The female has no wings, but crawls up the tree, and lays her eggs on the branches in May, in clusters of 60 to 100 in each, glued to each other and to the bark by a greyish varnish impervious to water; the little worms fall upon the leaves, and, when numerous, devour them all, leaving only the mid-ribs. They leave the trees when about four weeks old, and descend into the ground. Their effects are most visible in June, when the trees, divested of their foliage, appear as if scorched by fire.

As the female cannot fly, the great point is to prevent her from crawling up; for this purpose various means have been tried and are recommended. One of the most effectual is to tie strips of canvas around the tree and cover it with tar, renewing the tar during their whole season of rising, or from October till May. Another is, to make a close fitting collar of boards around the base of the tree, and keep them covered with tar. Mr. Jonathan Dennis, of Portsmouth, Rhode Island, obtained a patent for a circular leaden trough filled with oil, which proves an effectual preventive.

7. *Caterpillars.*—Of these there are many kinds that are more or less destructive to the foliage of fruit trees; but the Caterpillar, described by Professor Harris as the American Tent Caterpillar, is the one that commits such general and extensive devastation in our orchards, and especially in certain seasons. The moth deposits its eggs in July, in large rings, on the branches of the trees; these remain in that state until the following season, when they are hatched in the latter end of May or beginning of June. Each ring produces three or four hundred caterpillars, and these weave a sort of web to live in. The appearance of a tree with three or four of these tents upon it, and the leaves completely devoured, is really frightful. There are two ways of destroying them: one is, to examine

the trees carefully in February or March, at pruning time, and destroy the clusters of eggs by cutting off and burning the branches on which they are found. The next is to destroy the caterpillars in their tents after they are hatched. There are various ways of doing this, according to people's fancy and ingenuity. The quickest and most effectual method is to take a ladder, ascend the trees, and remove every nest with the hands. The early morning should be chosen, when they are in the nests. Some put a round brush on a pole and put it in the nests, and by giving it a few turns web and all are removed.

8. *The Cherry and Pear Slug.*—This is a most destructive insect. They appear in June and July for the first, and a second brood afterwards, small, slimy, dark-brown slugs on the upper surfaces of the leaves of the cherry and pear. They devour greedily the parenchyma of the leaves, leaving only the bare net-work of veins. In a short time growth is completely stopped.

Stocks for budding require careful watching, for a day or two of these slugs may prevent them from being worked that season. We destroy them by throwing fine earth taken up with the hand among the trees, and by ashes or slaked lime, when the earth is not sufficiently dry and fine. The caustic properties of lime and ashes render them more certainly destructive to the slug, and they should always be used in preference to common earth, where only a few trees are to be gone over.

A liberal syringing with the tobacco and soap liquid recommended for the aphis, but in a weaker state, is serviceable after the ashes and lime. It must be remembered that one application will seldom be sufficient. Some escape even to the third or fourth; but in all cases the warfare should be sustained whilst one remains. Like the aphis they are generally most troublesome in warm and dry seasons.

9. *The Curculio or Plum Weevil.*—This is a small greyish brown beetle nearly a quarter of an inch long; the wing covers form two little humps on the back, which give it a roundish appearance, and it has a long crooked snout, well adapted to its destructive propensities. They can fly, but are not active; and by jarring the part on which they stand, suddenly, they fall to the ground, draw in their legs and appear dead. It deposits its egg in a semicircular incision which it makes in the young fruit; it there hatches, eats into the fruit, and causes it to fall while yet green. It is the most troublesome of all insects injurious to fruits. In some places it destroys the entire crop of plums, apricots, and nectarines, and attacks even the cherry and the apple. The peach, even, is not wholly exempt, notwithstanding its coat of down. Almost every remedy that ingenuity can devise has been tried. This whole book would not contain what has been written on the subject in one year alone. Yet no complete, effectual remedy has been discovered. The strongest liquid applications of lime, soap, and tobacco—the most powerful and offensive odors, that repel any other insects, are entirely harmless and inoffensive to the curculio. There seem to be really but two means worthy of being resorted to. One is, to pave, or in some other way harden, the surface of the ground, so that the grubs cannot enter it to complete their transformations. This is found efficient where no other trees are in the immediate vicinity not paved. We have seen many instances where good crops were obtained by this mode. The fact that they are, as a general thing, less troublesome in stiff clay soils than in light porous ones, is alone a proof of the efficacy of a stiff or impenetrable surface soil.

Add to this the picking up of fruit containing the grub as soon as it drops from the tree, and before the worm has a chance to escape.

To accomplish both these ends, some people have planted their plums and apricots in a small enclosure by themselves, adjoining the hog-pen, and as soon as the fruits begin to drop, these animals are admitted, and gather all up, and, at the same time, tread the ground so firmly that it is almost as good as if it were paved.

This is probably the easiest and best way to ensure a crop of the fruits attacked by this insect.

Another way is, to jar the tree daily three or four times a day, from the moment they begin to appear, which is when the fruit is the size of a pea, until they have disappeared, or the fruit begins to ripen, when it is no longer attacked.

Serious injuries have been inflicted on plum trees, by thoughtlessly striking the bark of the trunk or a large branch with a mallet to jar the trees. The safer way is to strike on the end of a cut branch, or to fix a cushion of some soft material on the end of a short stick, and place the cushion on the tree, and strike the other end with the mallet. The insects are much easier jarred off in the cool of the morning while they are comparatively torpid.

Before commencing to jar them down, a white sheet or cloth, wide enough to cover all the ground under the branches, should be spread to receive the insects as they fall, so that they may be destroyed. This was recommended through the "Genesee Farmer," by David Thomas, twenty years ago.

From repeated observations, I am inclined to believe that it is quite sensitive to cold, for it is well known that in the cool of the morning it is always in a comparative state of torpor; and in the cold seasons of 1849–50, when our peach trees and fruit were so greatly injured, the curculio was driven off, and we had a most abundant crop of plums. A cold day or two may not affect it; but when it continues for two weeks, as in the years referred to, it seems to be rendered powerless for that season.

10. *Ants.*—These are not very destructive, yet they sometimes do considerable injury to beds of seedlings, by making their hillocks among them, and they also infest ripe fruits.

Boiling water, oil, or spirits of turpentine, poured on their hillocks, disperses them; and if wide-mouthed bottles, half filled with sweetened water or syrup, be hung among the branches of a tree when the fruit is attaining maturity, ants, wasps, flies, and beetles of all sorts that prey greedily upon sweets, will be attracted into them.

Mr. Downing, who recommends this as a "general extirpator suited to all situations," says, "that an acquaintance caught in this way, in one season, *more than three bushels* of insects of various kinds, and preserved his garden almost entirely against them."

A gentleman in Detroit, who was very careful of his garden, informed me that he had pursued this method of trapping insects with results that perfectly astonished him. He had to empty the bottles every few days to make room for more. A very good way of trapping and killing ants is, to besmear the inside of flower pots with molasses, and turn them on their mouths near the hillock; the insects will soon assemble inside on the molasses, when they are easily destroyed by a handful of burning straw.

11. *The Peach Tree Borer.*—This is a most destructive insect when allowed to increase for a few years without molestation. We have seen whole orchards of fine trees ruined by them. They sometimes attack even young trees in the nursery, and commit serious depredations on their collar, rendering them in many cases quite unfit for planting. Their multiplication should be prevented by all possible means. The eggs are deposited in summer

on the base of the trunk, near the collar, where the bark is soft. There they are hatched, and bore their way under the bark of the tree, either in the stem or root, or both, producing an effusion of gum. Where trees are already affected, the proper course is, to remove the earth from around the collar of the root, clean away the gum, destroy any cocoons that may be found, trace the grub through its holes in the tree, and kill it; then fill up around the tree with fresh earth, and place a shovelful or two of ashes around the base. One of the best orchards in the vicinity of Rochester was at one time nearly ruined by the prevalence of this grub, when it changed proprietors, and the present one adopted and followed the plan recommended above, until there is not the trace of one left. The ashes or slaked lime should be applied every spring, and at the end of summer may be scattered about the tree; both ashes and lime form an excellent dressing for the peach.

12. *The Rose Bug.*—The eggs of this insect are laid in the earth, where they are hatched, and from which the bug emerges about the rose season.

In some seasons and in some localities they appear like grasshoppers in vast multitudes, and commit extensive ravages, not only on the rose but fruit trees, and all other green things. There is no other way known to combat them, but to crush them with the hand—to spread cloths around the trees, and shake them down on it, and kill them. They are stupid, sluggish things, and fall as though they had no life.

In some cases fruit trees have been protected by covering them with millinet.

13. *Leaf Rollers.*—In May and June these insects may be found on the leaves of fruit trees, and especially on the pear; they form themselves a sort of cocoon out of the leaf. The leaves attacked by them should be removed and

destroyed, in order to prevent their increase. The eggs are deposited on the young leaves by some of the multitudes of spring beetles.

Section 3.—Animals Injurious to Fruits and Fruit Trees.

1. *Birds.*—As a general thing, birds are more the friends than the enemies of the garden. Many of them subsist in greater part on insects, and thus perform services that are by no means appreciated. The early cherries are generally the greatest sufferers by them, and various devices are practised to frighten them away, the most cruel of which is shooting. Moving objects resembling the human figure, bits of looking-glass or tin suspended among the branches, etc., are often effectual. Dwarf trees are easily covered with thin netting supported on poles and fastened at the base of the tree.

2. *Field Mice.*—The most effectual preventive is clean culture. Leave no grass, weeds, rubbish, or heaps of stones around the garden or orchard, and the mice will seldom be troublesome. Their operations of girdling are principally carried on beneath the snow, and when this is firmly trodden down as soon as it falls, it obstructs their way. A correspondent of the "Horticulturist" states that he has found tin tubes fixed around the base of the tree, an effectual remedy; and Mr. Hooker, of Rochester, has successfully driven them off with poison. He takes a block of wood six inches long and three or four square, and bores it lengthwise with an inch and half auger nearly through, and places in the lower end some corn meal and arsenic. He places these blocks among the trees, mouth inclined downwards, "to keep the powder dry."

3. *Moles.*—These are easily poisoned and driven off, by putting pills of flour mixed with arsenic into their holes, and shutting them up. We have seen them banished by bits of dried codfish placed in the entrance of their holes.

4. *Cats* often commit serious depredations on trees by scratching the bark. Quite recently we saw a large number of beautiful fruit trees nearly ruined by them. A few briers secured around trees in the vicinity of the house, where they frequent most, will be a sufficient protection.

5. *Hogs.*—It is not generally supposed that these animals will attack trees; but we have heard of a western farmer who turned in a large number of them to consume the corn that had been grown in his young orchard. When the corn began to grow scarce they attacked the trees, and not one out of several hundred but was completely girdled—the bark gnawed off as far up as the brutes could reach.

Where it may be desirable to turn hogs into an orchard, unless the feed be very abundant, the trees should be protected around the base with thorns, briers, or some prickly brush.

CHAPTER IV.

NURSERY, ORCHARD, AND FRUIT GARDEN IMPLEMENTS.

THE following are the principal implements used in the propagation, pruning, and cultivation of fruit trees:

SECTION 1.—IMPLEMENTS OF THE SOIL.

The Subsoil Plough is the great reformer of the day in the preparation of soils of all qualities and textures, for nursery, orchard, or garden trees. It follows the ordinary plough in the same furrow; and the largest size, No. 2, with a powerful team, can loosen the subsoil to the depth of eighteen inches. No. 1 will be sufficient in clear land when the subsoil is not very stiff.

The One-Horse Plough.—Similar to the common plough used by farmers. It is a labor-saving implement for cultivating the ground among nursery trees or orchards closely planted. The horse should be steady, the man careful, and the whiffle-tree as short as possible, that the trees need not be bruised. It should neither run so deep nor so near the trees as to injure the roots.

The Cultivator.—This with the plough obviates the necessity of spade-work, and, in a great measure, hoeing. If the ground be ploughed in the spring, and the cultivator passed over it once every week or two during the summer, all the hoeing necessary will be a narrow strip of a few inches on each side of the row. The double pointed steel-toothed, with a wheel in front, is the best.

The One-Horse Cart.—This is an indispensable machine in the nursery, orchard, or large garden. Four-wheeled wagons are difficult to unload, and require a great deal of space to turn in. The cart can be turned in a circle of twelve or fourteen feet, and the load discharged in a moment, simply by taking out the key that fastens the body to the shafts, throwing it up and moving the horse forward. Our carts are about six feet long and three wide in the body, shafts six feet long, wheels four and a half feet high, and tire two and a half wide to prevent them from sinking into the ground. The box is about a foot deep, and when large loads are to be carried a spreading board is put on the top with brackets. Cost from $30 to $50.

The Wheelbarrow (fig. 132).—Every man who has a rod of ground to cultivate should possess this machine. In small gardens it is sufficient for the conveyance of all manures, soils, products, etc., and in larger places it is al-

Fig. 132.
Wheelbarrow.

ways needed for use, where a cart cannot go. The handles or levers should be of ash or some tough wood, and the sides and bottom of any light wood. The wheel is soft wood, shod with iron.

Fig. 133.
Spade.

The Spade (fig. 133).—The best kinds of these in use

are Ames' cast-steel; excellent, strong, light articles. They work clean and bright as silver. There are several sizes. For heavy work, trenching, draining, raising trees, etc., the largest should be used.

Fig. 134.
Shovel.

The Shovel (fig. 134).—This is used in mixing, loading, and spreading composts and short manures. The blade should be of cast-steel.

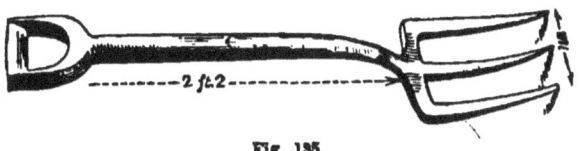

Fig. 135.
Digging Fork, or Forked Spade.

The Forked Spade (fig. 135).—This resembles a fork. It has three stout cast-steel tines, at least an inch wide, and pointed. It is used instead of a spade to loosen the earth about the roots of trees, to turn in manures, etc., being much less liable to cut and injure them than the spade.

Fig. 136.
Dung Fork.

The Dung Fork (fig. 136).—There are several kinds.

Those of cast-steel, cut out of a solid plate, with three or four tines, are the best, light and durable. It is the only implement proper for loading, mixing, or spreading fresh rough manures with facility and despatch.

The Pick.—This is a useful, and even indispensable implement in the deepening or trenching of soils with a hard subsoil that cannot be operated upon with the spade. It consists of an ash handle, and a head composed of two levers of iron pointed with steel, and an eye in the centre for the handle.

The Garden Line and Reel (fig. 137).— The line should be a good hemp cord, from one eighth to one fourth of an inch in diameter, attached to light iron stakes about eighteen inches long. On one of the stakes a reel is attached. This is turned by means of a handle, and the line neatly and quickly wound up.

Fig. 137.
Garden Line and Reel.

The Hoe.—This is a universal instrument in this country. In some cases, all the gardening operations are performed with it. Its uses in tree culture are to open trenches for seeds, to cover them, to loosen and clean the surface of the ground from weeds, &c. There are two kinds, the draw hoe, figs. 138, 139, 140, and the Dutch,

Fig. 138.
Square draw hoe.

Fig. 139.
Triangle draw hoe.

or thrust hoe, fig. 141; this we do not use at all. Of the

different kinds and forms of the draw hoe. The most generally useful is the square, a cast steel plate, about

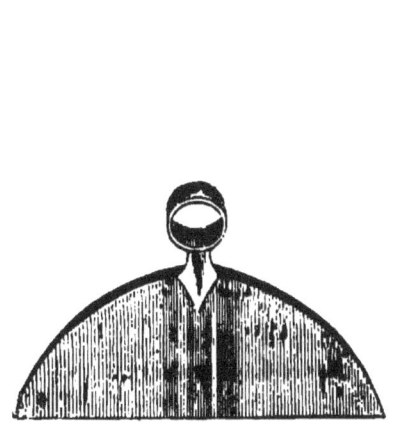

Fig. 140.
Semicircular draw hoe.

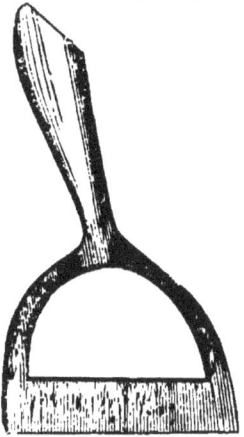

Fig. 141.
Dutch hoe.

six inches long and four wide, with a light smooth handle. The semicircular and triangle hoes may be advantageously used in certain cases.

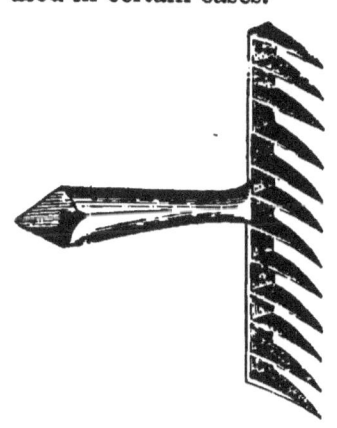

Fig. 142.
Garden rake.

The Rake, fig. 142, is used to level, smooth, pulverize, and clean the surface of the ground after it has been spaded or hoed, or to prepare it for seeds, &c. They are of different sizes, with from six to twelve teeth. The best are those of which the head and teeth are drawn out of a solid bar of steel. Those that are welded and riveted soon get out of order.

SECTION 2.—IMPLEMENTS FOR CUTTING.

The Pruning Saw.—This is used for cutting off

branches, either too large for the knife, or so situated that the knife cannot operate. It has various sizes and forms. Some are jointed, and fold like a pruning knife; others are like the common carpenter's handsaw, fig. 143, but smaller and stouter.

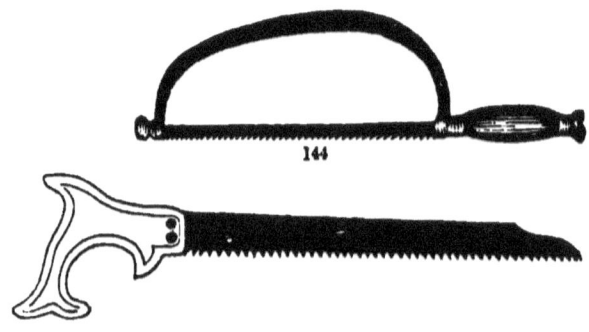

Fig. 143.—Pruning saw.   Fig. 144.—Bow saw.

The Bow Saw (fig. 144).—This is the most generally useful form for the gardener, or nurseryman. The blade is very narrow, and stiffened by an arched back. It is fastened at both ends by a rivet to the screw on which the back turns, and by which it is adapted to different purposes. It is indispensable in making horizontal cuts, close to the ground, as in heading down.

Some are set with a double row of teeth on one side, and the edge is much thicker than the back; these work much easier than those toothed in the ordinary way, and it would be an object to have them where much saw pruning is to be done. Wherever the saw is used, the cut surfaces should be pared smooth with the knife, to facilitate its healing.

Long handled pruning saws are sometimes recommended, but never should be used in pruning fruit trees. The branch to be operated should be reached by means of a ladder, if need be, within arm's length, and cut with a common saw.

Hand Pruning Shears (fig. 145).—There is a kind of these made now, that having a moving centre, as in the

Fig. 145.

figure, make a smooth *draw* cut almost equal to that of a knife, and it is a very expeditious instrument in the hand of a skilful workman. In pruning out small dead branches, shortening in peach trees, &c., it will perform four times as much work as a knife.

Pole Pruning Shears.—These resemble the hand shears, but are worked by a string passing over a pulley, and are fixed on a pole of any required length. They are used in cutting scions, diseased shoots, &c., from the heads of lofty standard trees.

Fig. 146.—Grape scissors.

Grape Scissors.—These are small sharp pointed scissors for thinning bunches of grapes.

The Pruning Knife.—The best for general purposes

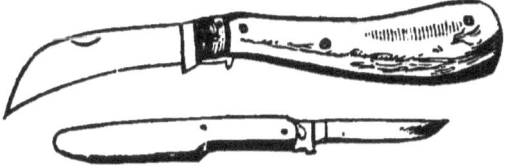

Fig. 147.—The pruning knife. Fig. 148.—The budding knife.

are those of medium size, with a handle about four inches long, smooth, slightly hollowed in the back; the blade about three and a half inches long, three-quarters of an inch wide, and nearly straight. For very heavy work a larger size may be necessary. "Saynor's" (English) knives of this kind are unsurpassed in material and finish.

They are to be had in the seed stores at $1 to $1 25 a-piece.

The Budding Knife.—This is much smaller than the pruning knife, with a thin straight blade, the edge sometimes rounded at the point. The handle is of bone or ivory, and has a thin wedge-shaped end for raising the bark. Budders have various fancies about shape and size.

Fig. 149.—Grafting Chisel.

The Grafting Chisel. —This is used for splitting large stocks; the blade is about two inches long, and an inch and a half wide, in the shape of a wedge; the edge curved so as to cut, and not tear the bark; the handle eight or ten inches long, at the end of which is a narrow wedge to keep the split open until the scion is inserted. The whole is of steel. Some are made with the blade in the middle, the wedge at one end, and a hook to hang it by on the other.

Foote's Stock Splitter.—This is an implement invented by A. Foote, Esq., of Williamstown, Mass., to facilitate cleft grafting. It consists of a sharp blade, *c*, and a groove,

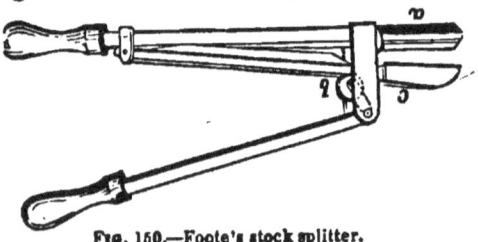

Fig. 150.—Foote's stock splitter.

a, sheathed with leather; the handles are of wood, and the whole implement about eighteen inches long. The stock is placed in the groove, and the blade brought down upon it by the lever which acts upon a small wheel, *b*.

It performs splitting both neatly and quickly.

Section 3.—Ladders and Fruit Gatherers.

Ladders.—Of these there are many kinds. For the fruit garden, where the trees are low, the self-supporting ladder (fig. 151) is the most convenient and best. It should be made of light wood, with flat steps, so that a person can stand upon them and work. The back, or supports, consist of one or two light pieces of timber, fixed at the top with hooks and straps so as to be contracted or extended at pleasure. A ladder of this kind, six or eight feet high, will answer all the demands of a garden.

Fig. 151.
Self-supporting ladder

Orchard Ladders are of various kinds. For pruning or gathering the fruit from lofty trees, a great length of ladder is necessary; it is therefore desirable that the material be as light as possible consistent with the necessary strength.

Sometimes these long ladders are composed of several smaller ones, that fit into one another, all mounted on a frame with a small wheel, by which they are easily moved about.

The Folding Ladder is a very neat and convenient article for many purposes. The inside of the styles is hollowed out, and the steps are fastened to them by means of iron pins, on which they turn as on hinges, so that the two sides can be brought together, the steps turning into the grooves or hollows in them, the whole appearing like a round pole, *B*. It is more easily carried and placed where wanted than the ordinary ladder. *A* represents it open, and *B* closed (fig. 152).

There are also self-supporting orchard ladders, composed of three upright pieces of any required length, and spread widely at the bottom to give them stability. Two

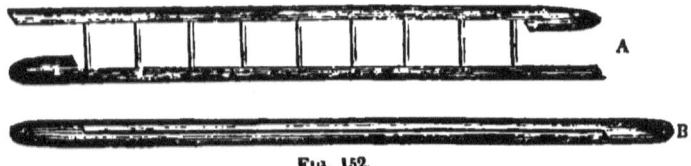

Fig. 152.
Folding ladder.

of the sides are fixed, and are furnished with steps all the way up. The third is longer and movable, and can be extended or contracted at pleasure.

A piece of board wide enough to stand upon can be extended from one side to the other, resting upon the steps at whatever height it is desirable to work. On the movable side a pulley is fixed, by which the baskets of fruit are let down as they are gathered. Two persons or more can ascend and work on a ladder at the same time. Fig. 153 represents one of these; *a*, *b*, the two fixed sides; *C*, the movable one. It is considerably used in France.

Fig 153.
French self-supporting orchard ladder.

The Orchardists' Hook.—Is a light rod, with a hook on one end, and a movable piece of wood that slides along it.

The person gathering fruit draws the branch towards him with the hooked end, and retains it there by means of the sliding piece which is hooked to another branch. This is an indispensable instrument in gathering fruit from large trees.

Fruit Gatherers.—Of these there are many designs by

which the fruit may be taken from the tree by a person standing on the ground. None of them are applicable to the gathering of fruits that are to be kept long, because it is impossible to avoid bruising them more or less, and besides this they operate slowly. They answer very well for gathering a few ripe specimens for immediate consumption. The *ladder*, *hook*, and *hand*, are the only safe and expeditious fruit gatherers. Some are made in the form of a vase of wood or tin placed on the end of a pole. The edge of the vase is toothed, and when the stem of the fruit is taken between two of the teeth, and slightly twisted, it drops. Others are composed of a pair of shears on the end of a pole, to which a basket is attached that slides up and down the handle.

The Grape Gatherer resembles a pair of shears combining the property of pincers. They cut a bunch of grapes, and hold it firmly until it is brought down. These are very useful for gathering a few bunches of grapes from the top of a house or trellis (fig. 154).

FIG. 154.
Grape Gatherer.

SECTION 4.—MACHINES FOR WATERING.

The Hand Syringe (fig. 155).—This is a very useful implement for sprinkling and washing the foliage of trees in dry weather. There are various kinds made of tin, copper, and brass, and sold at various prices. Whatever sort is used should have several caps (*A*) to regulate the quantity or shower of water discharged; and they should also have an inverted or "gooseneck" one (*B*) to throw

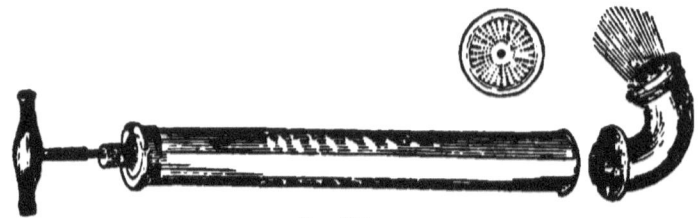

Fig. 155.
Hand Syringe.

the stream, if necessary, on the under side of leaves, or in any oblique direction.

There are, also, *hand engines, barrel engines,* and *barrow engines,* all of which are very useful. In every large garden there should be both the syringe and one of these engines; for watering is a most important affair in gardening under our hot sun and protracted droughts.

Fig. 156.
Barrow Engine.

The Barrow Engine (fig. 156) is the most useful for general purposes; it is easily moved from one place to another. The improved kinds are easily worked, and the water-box being provided with a strainer, excludes anything likely to derange its operations.

The Garden Watering-pot (fig. 157).—This is a tin or copper vessel that may hold from one to four gallons of water, with a spout six or eight inches long, by which the water is discharged. There should be a rose or roses, as in cut, to fit on the spout, pierced with *large* or *small* holes, by which the water can be discharged in a shower. Every pot may have several roses pierced with holes of various sizes, to adapt them to different purposes.

Fig. 157.
Garden Watering-pot.

INDEX.

Air, importance of, to the germination of seeds, 46.
Alluvial soils, nature of, 49.
Almonds, select varieties of, 351.
Anthers, the, 28.
Ants, method of destroying, 373.
Aphis, the, how to destroy, 134, 366; the woolly aphis or American blight, 366; the scaly aphis or bark louse, 367.
Apple, the, principal stocks in use for, 108; time for budding, 133; dwarf apple tree, 189; pruning the, 203; management of the standard apple tree, 204; dwarf standards, 206; pyramids, 207; pruning the branched yearling, 210; treatment of two-year old nursery trees, 211; management of the fruit branches, 217; pruning and management of, as a dwarf on the paradise stock, 218; as an espalier, 220; renovation of pyramidal trees, 271; fire blight of, 361; insects infesting the, 367; apple tree borer, description of, and means of destroying, *ib.*; the apple worm, 368.
Apples, forms of, 40; abridged descriptions of one hundred and thirty-three select varieties of summer, autumn, winter, and apples for ornament and preserving, 279-297; small select lists of, suitable for Western New York, 297.
Apricot, the, stocks for, 119; as a dwarf standard, 194; pruning and management of, 245.
Apricots, abridged descriptions of seven select varieties of, 320, 321.
Ash, the mountain, as a stock for the pear, 115.

Bark, the outer, 4; the inner, 5.
Barrow engine, the, 388.
Berberries, method of propagating, 148; select varieties of, 347.
Birds, manner of protecting fruit against, 375.
Black knot, the, or plum wart, 364.
Blackberries, 347.
Blossoming, period of, influenced by various conditions, 31, 32; in alternate years, 34.

Branches, the, various subdivisions of, 7; pruning, 88; removal of large, 89; management of fruit, 217.

Budding, method of, 70; conditions necessary to the operation of, 71; implements requisite for, 71, 72; cutting, preparing, and preserving the buds, 72; chief difficulty experienced by beginners in, 74, 75; time for, 133; insertion of the bud, 134; untying the buds, 135; treatment of the growing bud, 136.

Buds, nature and functions of, 14; different names and characters of, 15; variations in the size, form, and prominence of leaf buds, 17; fruit buds 18; leaf and fruit buds how distinguished, 20.

Bushes, dwarf, 142.

Calcareous or chalky soils, nature of, 49.
Calyx, the, 28.
Cambium, nature of, 24.
Canada or wild plum, the, 120; time for budding, 133.
Canker worm, the, 368.
Cart, the one-horse, 378.
Caterpillars, methods of destroying, 369.
Cats, method of protecting trees from, 376.
Cherries, forms of, 43; abridged descriptions of fifty-five select varieties of heart, bigarreau, duke and morello, and new and rare cherries recently brought to notice, 321–329; small select lists of, 329.
Cherry, the, principal stocks in use for, 115; time for budding, 133; in the pyramidal form, 191; pruning the, 224; as a standard, 225; to form a round open head, 225; as a pyramid, 226; as an espalier, 227; as a dwarf or bush, 228; effects of the disease called gum on, 362.
Chestnuts, method of propagating, 148; select varieties of, 351, 352.
Chinese layering, description of the process so called, 122, 125.
Chisel, the grafting, 384.
Clayey soils, nature of, 48.
Cleft grafting, 79.
Corolla, the, 28.
Cotyledons, the, 44.
Cultivator, the, 377.
Curl of the leaf in the peach, the, 364.
Currant, the, method of propagating, 147; pruning and management of, 265, as a pyramid and espalier, 266.
Currants, abridged descriptions of eleven select varieties of, 341, 342.
Cutting back, object of, 207; process of, 208; summer management of trees cut back, 213.
Cuttings of fruit trees, how to make, 65; time of making, soil proper for, and time of planting, 66; method of preserving, 67; treatment of, when transplanted, 127.

Diseases of fruit trees, description of and remedies for, 361.
Distance at which standards should be planted in the nursery, 129.
Double-working, explanation of, 81.
Doucain, the, method of propagating, 110.
Draining, process of, 52.
Dubreuil, M., his summary of the general principles of pruning, 96.
Dung fork, the, 379.
Dwarf standards, 131; management of, 140, 206; dwarf bushes, 142; dwarf apple tree, 189; the cherry as a dwarf or bush, 228.

Enclosures, various kinds of, for orchards, 163: for fruit gardens, 181.
Espaliers, method of forming, 143; proper distances between, 201; the apple and pear as, 220; the cherry as, 227; the peach as, 236; method of laying in, and fastening to walls and trellises, 243; the currant as, 266.

Fences, materials for making, for orchards, 163; for gardens, 181.
Fibres, the, or rootlets, 2.
Fig, the, propagation of, 160; suitable soil for, 261; pruning and training of, *ib.*
Figs, select varieties of, 350, 351.
Filbert, the, method of propagating, 148; as a dwarf and pyramid, 194; pruning and training, 257; account of the management of filbert orchards in the county of Kent, England, *ib.*
Filberts, abridged descriptions of six varieties of, 352.
Fire blight, the, of the pear, apple, and quince, 361.
Flowers, different parts of, 28; sexual distinctions of, 29; method of impregnation of, 30; double, 31; different characters of, 32; hybridization of, 33.
Foote's stock splitter, 384.
Fruit rooms, requisites for, 359.
Fruit trees, names, descriptions, and offices of the different parts of, 1; fruit branches and fruit spurs, 10; fruit buds, 20; propagation of, by seeds, 60; general carelessness in the selection of the seeds of, 61; production of new varieties of, 63; propagation of, by cuttings, 65; propagation of, by layering, 67—by suckers and by budding, 70; propagation of, by grafting, 75; pruning of, 83; method of regulating the growth of, 92; method of promoting the fruitfulness of, by pinching, 94; budding, grafting, and management of, in the nursery, 132; taking up, from the nursery, 150; method of packing, 151; process of heeling in, 153; permanent plantations of, 157; proper soils for different, 162; points to be considered in selecting varieties of, for an orchard, 164; arrangement of, in an orchard, 167; pruning and preparing, for planting, 174; how to select, for the fruit garden, 188, 195; age of, for the fruit garden, 195; arrangement of, in the fruit garden, 190; sundry operations connected with the culture of, 272; diseases peculiar to, 361; insects injurious to, 365.

Fruitfulness, method of promoting, by pinching, 94.
Fruits, nature and classification of, 34, 35; forms and colors of, 36; different parts of, *ib* ; size of, circumstances influencing the, 37; classification of the size of, 39; form of, 40; color and flavor of, 43; abridged descriptions of select varieties of, 277; directions for the gathering, packing, transportation, and preservation of, 354; the best fruit gatherers, 387.

Fruit Garden, the, general remarks on, 178; situation for, 179; soil of, 180; enclosures for, 181; manner of laying out, 183; the mixed, or fruit and kitchen, 185; walks in, 186; a supply of water important for, 187; how to select trees for, 188, 195; age of trees for, 195; arrangement of trees in, 199; implements used in, 377; machines for watering, 387.

Gathering fruits, proper period for, 354; mode of, 355; disposition of fruits after gathering, 356; implements used in, 386.
Germination, process of, 45.
Gooseberries, abridged descriptions of five select varieties of, 342, 343.
Gooseberry, the method of propagating, 147; pruning, 262; method practised in Lancashire to produce large gooseberries, 264; severely affected by the mildew, 363.
Grafting, process and objects of, 75; implements used in, and grafting composition, 77; whip-grafting on the root, 78; cleft grafting, 79; precautions to be taken in, 81; double-working, *ib.*; implements used in, 384.
Grape vine, the, methods of propagating, 145; general observations on the management of, 245; planting, 248; pruning, 249.
Grapes, culture of foreign, in cold vineries, 253; abridged descriptions of four select varieties of hardy grapes and nine varieties of foreign grapes, 348-350; instrument for gathering, 387.
Gravelly soils, nature of 49.
Growth of trees, method of regulating the, 93.
Gum, the, in stone fruits, 362.

Heading down, process of, 85.
Heart or perfect wood, the, 5.
Heat, effect of, on the germination of seeds, 46.
Heeling in, process of, 153.
Hoe, the, two kinds of, 380.
Hogs, method of protecting trees against injuries caused by, 376.
Hook, the orchardist's, 386
Horse plum, the, 119.
Hybridization, explanation of the process of, 33.

Implements used in the orchard, nursery, and fruit garden, **377**.
Insects, the principal, which infest fruit trees, 365.

Knife, the pruning, 383; the budding, 384.

INDEX. 395

Labels for trees in the nursery, 149; manner of labelling, 150.
Ladders, orchard, folding, and self-supporting, 385.
Layering, process of, 67, 125; propagation of plums by, 122; treatment of layers when transplanted, 127.
Leaf rollers, 374
Leaves, structure and functions of, 21; different forms and characters of, 25.
Light, exclusion of, necessary for the germination of seeds, 47.
Line and reel, the, 380.
Loamy soils, nature of, 49.

Mahaleb, the, 117.
Manures, importance of, 54; preparation of, 55; special, 56; modes of applying, 58; liquid, ib.
Manuring, proper method of, 272.
Mazzard seedlings, 115.
Medlar, the, treatment of, same as that of the quince, 224.
Medullary rays, 7.
Mice, method of protecting fruit trees from the ravages of, 375.
Mildew, the, in the peach, 363.
Moisture, effects of, on the germination of seeds, 45.
Moles, method of guarding against the ravages of, 376.
Mulberries, method of propagating, 148; varieties of, 348
Mulching, operation of, 176, 273.

Nectarine, the, stocks for, 119; as a dwarf standard, 194; pruning the, 246.
Nectarines, abridged descriptions of seven select varieties of, 329, 330.
Nursery, the, soil of, 105; method of laying out, 107; situation of, and succession of crops in, ib.; directions for planting stocks in the nursery rows, 127; budding, grafting, and management of trees in, 132; treatment of the soil in, 144; labels for trees in, 149; taking up trees from, 150; implements used in, 377.

Orchard, the, situation of, 158; soil of, 161; how to prepare the soil for, 162; selection of varieties of fruit trees for the family, 164; kind of trees to be selected for, 166; arrangement of the trees in, 167; selection of trees for the market, 169; planting the, 175; management of trees in, 177; implements used in, 377, 385.
Ovary, the, 29.

Packing trees, proper method of, 151; method of packing fruits, 357.
Paradise, the, method of propagating, 111.
Parenchyma, nature of, 22.
Parsons, S. B., his orchard of pear trees, 172.
Peach, the, principal stocks for, 117; times for budding, 133; as a dwarf standard, 193; pruning and management of, 229; to form the head of a

standard peach tree, 231; root pruning, 233; conducted in the form of a vase, *ib.*; as an espalier, 236; symptoms of the yellows and mildew in, 363; insects infesting, 373.

Peaches, forms of, 42; abridged descriptions of thirty-eight select varieties of freestone and clingstone, 330–335; select list of, 335.

Pear, the, principal stocks in use for, 111; time for budding, 133; as a pyramid, 190, 207; in the dwarf standard form, 191, 206; management of the standard pear tree, 204; pruning the branched yearling, 210; treatment of two-year old nursery trees, 211; management of the fruit branches, 217; as an espalier, 220; renovation of pyramidal trees, 271; fire blight of, 361; pear leaf blight, 362.

Pears, forms of, 41; varieties of, that succeed well on the quince, 172; abridged descriptions of one hundred and eighty-two select varieties of summer, autumn, winter, pears for baking and stewing, and new and rare varieties, 299–319; select assortments of, 312; method of packing for distant markets, 358.

Peaty soil, nature of, 49.

Petals, the, 28.

Pick, the, 380.

Pinching, nature and objects of, 92

Pistil, the, 29.

Pith, the, 5.

Plant louse, the, ravages of, and method of destroying, 365, 366.

Plantations of fruit trees, different kinds of, 157.

Plants, exhalation of moisture and gases by, 22; propagation of, 60.

Plough, the subsoil and one-horse, 377.

Ploughing, subsoil, 50.

Plum, the, stocks for, 119: the horse, *ib.*; the Canada, or wild, 120; the cherry plum, *ib.*; the sloe as a stock for, 121; propagated by layers, 122; time for budding, 133; as a pyramid, 192; as a dwarf standard, 193; pruning and management of, 244; diseases of, 364.

Plums, forms of, 42; abridged descriptions of fifty-one select varieties of, 335–340; small select lists of, 341.

Plumule, the, 44.

Pruning, importance of the operation of, 83; various objects to be attained by, 84; to direct the growth from one part to another, *ib.*; heading down, 85; to maintain an equal growth, to renew growth, and to induce fruitfulness, 86; pruning the roots and pruning at the time of transplanting, 87; mechanically considered, 88; season for, 91; general principles of, as laid down by Dubreuil, 96; directions for the pruning of stocks, 125; pruning the apple and the pear, 203; the quince, 222; the cherry, 224; the peach, 229; the plum, 244; the apricot, 245; the nectarine, 246; grape vines, *ib.*; the filbert, 257; the fig, 260; the gooseberry, 262; the currant, 265; the raspberry, 267; implements used in, 381.

Pyramids, management of, 140, 207; renovation of, 271.

Quince, the, as a stock for the pear, 113; erroneous ideas concerning, 114; as a dwarf and pyramid, 194; pruning and training of, 222; fire blight of, 361.
Quinces, abridged descriptions of seven select varieties of, 319, 320.

Rake, the, 381.
Raspberries, forms of, 43; method of propagating, 148; abridged descriptions of six select varieties of, 343, 344.
Raspberry, the, planting of, 267; pruning, 268; manuring and training, 269; French and English modes of training, 269, 270.
Root, the, the several parts of, 2; growth of, 3; whip-grafting on, 78 pruning, 87, 90; method of planting root-grafts, 131.
Rosebug, the, 374.

Sandy soils, nature of, 48.
Sap, the, ascent, assimilation, and descent of, 24; tendency of, to the growing points at the top of a tree, 85.
Sap-wood, the, 5.
Saw, the pruning, 381; the bow, 382.
Scions, selection and treatment of, for grafting. 76.
Scissors, grape, 383.
Seed. the, composition of, 44; germination of, 45; propagation by, 60; selection of, 61.
Seedling apple, the common or free stock, preparing, saving, and planting the seed of, 108; after management of, 109; the pear seedling, 111.
Shears, hand pruning and pole pruning, 383.
Shovel, the, 379.
Sloe, the, as a stock for the plum, 121
Soils, different kinds of, 48; different modes of improving, 50; proper, for the orchard, 161; annual cultivation of the soil, 272.
Spade, the, 378; the forked spade, 379.
Spongioles, the, 3.
Staking, process of, 176.
Stamens, the, 28.
Standards, management of, 137; dwarf, management of, 140.
Stem, the, the different parts of, 4; structure and growth of, 6; the branches divisions of, 7; pruning, 88.
Stigma, the, 29.
Stocks, necessity of a close alliance between, and grafts, 76; description and propagation of, 108; for the apple, ib.; for the pear, 111; for the cherry, 115; for the peach, 117; for the apricot, nectarine, and plum, 119; transplanting, 122; time and manner of taking up, 124; pruning stocks, 125·

planting in the nursery rows, 127; treatment of, after planting, 132; time for budding, 133; preparation of, and insertion of the bud, 134.

Strawberries, forms of, 43; method of propagating, 147; abridged descriptions of twenty-six select varieties of, 344–346.

Style, the, 29.

Subsoil ploughing, 50; the subsoil plough, 377.

Suckers, propagation of fruit trees by means of, 70.

Syringe, the hand, 387.

Temperature, method of protecting trees against extremes of, 273.

Thorn, the, as a stock for the pear, 115.

Transplanting stocks, directions for, 122.

Tree, a, general remarks upon the structure of, 1; the root, 2; the stem, 4 the branches, 7; the buds, 14; the leaves, 21: the flowers, 28; the fruit 34; the seed, 44.

Trellises, form and construction of, 182.

Trenching, process of, 51.

Varieties of fruits, abridged descriptions of select, 277.

Vineyards, culture of, 252.

Walks, manner of laying out, in the fruit garden, 186.

Walnuts, method of propagating, 149; abridged descriptions of varieties of, 353.

Water, a supply of, important for a fruit garden, 187.

Watering, beneficial effects of, in fruit trees, 273; machines for, 387; the garden watering-pot, 389.

Wheelbarrow, the, 378.

Whip-grafting on the root, 78.

Wilder, M. P., compost recommended by, for gardening purposes, 57.

Yellows, the, in the peach, 363.

THE LIFE
OF
GEORGE WASHINGTON,
First President of the United States,
BY JARED SPARKS, LL. D.

NEW AND FINE EDITION, TWO VOLUMES IN ONE.

With Portrait, 674 pp. 12mo., Muslin, Price $1 25

"Let every mother's daughter's son
Be taught the deeds of Washington."

The materials for this volume have been drawn from a great variety of sources; from the manuscripts at Mount Vernon, papers in the public offices of London, Paris, Washington, and all the old Thirteen States; and also from the private papers of many of the principal leaders in the Revolution. The entire mass of manuscripts left by General Washington, consisting of more than two hundred folio volumes, was in the author's hands ten years. From these materials it was his aim to select and combine the most important facts, tending to exhibit in their true light the character of Washington.

☞ This is the most full, accurate and interesting biography of the Father of his Country, ever published, and is furnished at a very low price. Every American family should possess, and every American youth should read it.

THE LIFE OF LOUIS KOSSUTH,
Governor of Hungary;

INCLUDING NOTICES OF THE MEN AND SCENES OF THE HUNGARIAN REVOLUTION TO WHICH IS ADDED AN APPENDIX, CONTAINING HIS PRINCIPAL SPEECHES.

BY P. C. HEADLEY.

WITH AN INTRODUCTION, BY HORACE GREELEY.

One Volume, 461 pp. 12mo., Steel Portrait, Muslin, Price $1,25.

Mr. Headley has written an excellent memoir of Kossuth. It details, in easy, perspicuous narrative, the principal events of his life, bringing down the history almost to the present hour.—*N. C. Advocate.*

Sold by all Booksellers. Mailed, *post-paid*, to any address, upon receipt of price.

C. M. SAXTON, MILLER & Co., Publishers,
20 Park Row, New York.

Border Wars of the West,

COMPRISING THE

FRONTIER WARS

OF

PENNSYLVANIA, VIRGINIA, KENTUCKY, OHIO, INDIANA,

Illinois, Tennessee, and Wisconsin;

AND EMBRACING THE

INDIVIDUAL ADVENTURES AMONG THE INDIANS,

AND

Exploits of Boone, Kenton, Clarke, Logan,

AND OTHER

BORDER HEROES OF THE WEST

BY PROFESSOR FROST.

608 pp. 8vo., 300 Illustrations. Cloth $2 00. Morocco $2 50

The wars between the early settlers on the western frontier of our country and its aboriginal inhabitants, form an extremely interesting portion of history. The long period of time through which these wars extend, the large number of actions which they embrace, the variety of adventures and instances of individual heroism which they display, and the magnitude and importance of the territorial acquisitions in which they resulted, fully entitle them to form a separate history.

The Border Wars of the West, when we enter into their details, as gathered from the traditions received from those who were engaged in them, abound with interesting displays of human character. In them we may study the traits of the Indians, the terrible enemies of our forefathers, ancient possessors of the soil, who resisted their gradual but certain encroachments, with all the violence of savage fury, and all the stratagems of barbarous subtilty and cunning. Here, too, we may learn many useful lessons from the traits of character exhibited by the border settlers, exposed by their position to all the horrors of the midnight surprise with the dreadful accompaniments of the warwhoop, the massacre, the burning and plundering, murder and scalping, and followed by the weary sorrows of Indian captivity. We can never cease to admire the courage and fortitude with which the old border heroes and their not less heroic wives, confronted the dangers of a life on the frontier, and the activity, promptness, and determination with which they met and punished every assault.

Sold by all Booksellers. Mailed, *post-paid*, to any address, upon receipt of price

C. M. SAXTON, MILLER & Co., Publishers,

25 Park Row, New York.

ALL THE ARCTIC EXPEDITIONS

ARCTIC
EXPLORATIONS AND DISCOVERIES
DURING THE NINETEENTH CENTURY.
BEING DETAILED ACCOUNTS OF
THE SEVERAL EXPEDITIONS TO THE NORTH SEAS,
BOTH ENGLISH AND AMERICAN, CONDUCTED BY
ROSS, PARRY, BACK, FRANKLIN, M'CLURE AND OTHERS.
INCLUDING THE FIRST GRINNELL EXPEDITION,
UNDER LIEUTENANT DE HAVEN, AND THE
FINAL EFFORT OF DR. E. K. KANE
IN SEARCH OF SIR JOHN FRANKLIN.
BY SAMUEL M. SMUCKER.
517 pages 12mo., with Illustrations. Price $1 00.

Repeated, bold, and daring enterprises have been undertaken during the present century, to explore the hidden recesses of the Northern seas, the vast frozen region of everlasting snow, of stupendous icebergs, of hyberborean storms, of the long, cheerless nights of the Arctic Zone. To navigate and explore these dismal realms, men of extreme daring, of sublime fortitude, of unconquerable perseverance, were absolutely necessary. And such men possessed one great element of distinguishing greatness, of which the explorerers of more genial and inviting climes were destitute. Their investigations were made entirely without the prospect of rich reward, and chiefly for the promotion of the magnificent ends of science. The discovery of a north-western passage was indeed not forgotten; but it must be conceded that other less mercenary and more philanthropic motives have given rise to the larger portion of the expeditions which, during the progress of the nineteenth century, have invaded the cheerless solitudes of that dangerous and repulsive portion of the globe.

This work contains a narrative of the chief adventures and discoveries of Arctic explorers during this century. No expedition of any importance has been omitted; and the work has been brought down in its details to the present time, so as to include a satisfactory account of the labors, sufferings, and triumphs of that prince of Arctic explorers and philanthropists, Dr. Kane; whose adventures, and whose able narrative of them, entitle him to fadeless celebrity, both as a hero in the field, and as a man of high genius and scholarship.

The great chapter of Arctic adventure and discovery, may now be considered as closed. A concise and reliable account threfore, of those adventures and discoveries, with the novel and thrilling incidents attending them, cannot be otherwise than interesting in themselves, and acceptable to the public.

Sold by all Booksellers Mailed, *post-paid*, to any address, upon receipt of price.

C. M. SAXTON, MILLER & Co., Publishers,
25 Park Row, New York.

Books for Young Men.

Young Man's Book.

Or, Self-Education, by Rev. WM. HOSMER, frontispiece on Steel, 291 pp. 12mo., Muslin, 75 cents.

A constant and persevering determination to tread in the steps which naturally lead to honor.—*J. Q. Adams.*

We are glad to see such books launched out upon the sea of public mind. They are safe crafts and freighted with wholesome and salutary truths. Mr. Hosmer writes for the good of his readers, and deals in the plain and practical matters of man's education and course of life. He is a safe counselor for the young, and never gilds a pernicious sentiment in the seductive drapery of rhetoric. This book has the calm and truthful characteristics of the author. He writes with an eye upon man's highest destiny upon earth, and the great results of his living, in Heaven.—*Cayuga Chief.*

Gift Book for Young Men.

Or Familiar Letters on Self-knowledge, Self-education, Female Society, Marriage, &c., by Dr. WM. A. ALCOTT, frontispiece, 312 pp. 12mo., Muslin, 75 cents.

"Up! It is a glorious era!
Never yet has dawned its peer!
Up and work! and then a nobler
In the future shall appear!"

CONTENTS—LEADING HEADS.

Preliminary Remarks; Self-respect and Self-reverence; Self-knowledge; Self-dependence; Self-education; Harmony of Character; Self-instruction; Light Reading; Correct Conversation; The Schools; The Love and Spirit of Progress; Love of Inquiry; or, Free-thinking; Right Use of Ourselves; Physiology; Phrenology; Physiognomy; Traveling; Conscientiousness; Love of Excitement; On Purity; Models and Model Character; Decision and Firmness; Setting Up in Business; Money-getting; Pleasure Seeking; Mental Excitants; Respect for Age; Duties to the Aged; Politics and Political Duties; Female Society; General Duty of Marriage; Religion and Skepticism; The Christian Religion; Death and Futurity.

Golden Steps for the Young.

To Usefulness, Respectability and Happiness, by JOHN MATHER AUSTIN, author of "Voice to Youth," frontispiece on Steel, 243 pp. 12mo., Muslin, 75 cents.

We honor the heart of the writer of this volume as well as his head. He has here addressed an earnest and manly appeal to the young, every page of which proves his sincerity and his desire for their welfare. The subjects treated of in the different lectures are those indicated on the title page. Integrity and virtue, usefulness, truth and honor, are the "Golden Steps" by which the young may ascend to respectability, usefulness, and happiness.—*Albany State Register.*

Sold by all Booksellers. Mailed, *post-paid*, to any address, upon receipt of price.

C. M. SAXTON, MILLER & Co., Publishers,

25 *Park Row, New York.*

LIFE OF NAPOLEON BONAPARTE,
Emperor of France.

BY J. G. LOCKHART.

With Steel Portrait, 392 pp. 12mo., Muslin. Price $1 25.

> "The lightning may flash and the loud thunder rattle,
> He heeds not, he hears not, he's free from all pain;
> He has slept his last sleep, he has fought his last battle,
> No sound can awake him to glory again."

He was the greatest actor the world has known since the time of Cæsar. He sported with crowns and sceptres as the baubles of a child. He rode triumphantly to power over the ruins of the thrones with which he strewed his pathway. Vast armies melted away like wax before him. He moved over the earth as a meteor traverses the sky, astonishing and startling all by the suddenness and brilliancy of his career. Here was his greatness. The earth will feel his power till its last cycle shall have been run.

THE LIFE
OF
THE EMPRESS JOSEPHINE,
First Wife of Napoleon.

BY P. C. HEADLEY.

With Steel Portrait, 383 pp., 12mo., Muslin. Price $1 25.

> "Like the lily,
> That once was mistress of the field and flowers here,
> I'll hang my head and perish."

Josephine, for the times in which she lived, was a model of female character; and if this volume shall make the study of it more general, it will so far extend the admiration of the pure and beautiful, in contrast with all the forms of corruption humanity could present in a period of bloody Revolution. The Empress was a greater personage than Napoleon in the elements of *moral* grandeur, and retained her sovereignty in the *hearts* of the people, while he ruled by the unrivaled splendor of his genius.

Sold by all Booksellers. Mailed, *post-paid*, to any address, upon receipt of price.

C. M. SAXTON, MILLER & Co., Pub'lshers,

25 *Park Row* New York.

A Gem for Ladies!

"Instruction and amusement may combine,
As heat is blended with the beams that shine."

THE CHRISTIAN VIRTUES,

PERSONIFIED AND EXHIBITED AS A DIVINE FAMILY,

IN THEIR

DISTINCTIVE CHARACTERS, ASSOCIATIONS, MISSIONS, LABORS, TRANSFORMATIONS, AND ULTIMATE REWARDS.

AN ILLUSTRATED ALLEGORY.

BY REV. D. D. BUCK.

In One Volume, 290 pp. 12mo. Price $1 00.

OPINIONS OF THE PRESS.

From a Western Poetess.—The language is chaste and appropriate, the characters finely drawn, and the whole makes a fine family picture, which some great artist like Powers, or Durand, might chisel into palpable form, or cause to glow upon the almost breathing canvass with admirable effect.—Mrs. Jane Maria Mead.

The author evinces ingenuity in the construction of his Allegory, the perusal of which cannot fail to contribute to the amusement and instruction of the intelligent and pure-minded, to whom the volume is affectionately and respectfully dedicated. It is a well printed volume, &c., &c.—Rev. Dr. Floy—*National Magazine.*

The author of this work is already known to our readers by sundry contributions to our columns. He is also known to the public as the author of "Our Lord's Great Prophecy." In the present volume he has given us an illustrated Allegory—exhibiting after the manner of the Pilgrim's Progress, the various offices of the several Christian Virtues—their vocation *here*, their reward *hereafter*. The work exhibits not a little ingenuity in its conception as well as execution. It is quite readable, and will be equally instructive to the young.—Rev. D. W. Clark, D. D.—*The Ladies' Repository.*

The author has well and admirably combined instruction and amusement in his allegory of the "Christian Virtues." We commend the book strongly, and hope it may enter many a home—be a welcome guest at many a hearthstone.—*Literary Journal.*

This work, with just discrimination, paints all the christian virtues, and the attention is called, successively, to the origin, character, office &c. of all the members of the Divine Family, Religion, Truth, Liberty, Faith, Hope, Charity, Mercy, Justice, Contemplation, Impulse, Zeal, Industry, Patience, Humility, Virtue, Temperance. Of this family of sons and daughters, Religion is represented as the Divine Mother. The study of the book can but be a great aid to pious meditation.—*Christian Messenger.*

The excellence and beauty of the Christian Virtues, as here presented, cannot fail to commend them to the readers' approval, and to make a salutary and lasting impression upon their minds; while the intrinsic interest of the work, will delight, instruct, and improve.—*Syracuse Journal.*

The truthfulness of its characters, in all their movements and relations, challenges universal recognition—far more so than had they been drawn from individual, rather than universal experience. The author of the volume before us has taken hold of his subject with much of Bunyan's spirit and power, and in treating the great family of Christian virtues—Truth, Hope, Charity, Faith, Patience, &c.—in their separate or associated development in human life, he has invested them with more than mortal ability, and clothed them about with transcendent grace and beauty.—*Evening Mirror.*

Sold by all Booksellers. Mailed, *post-paid*, to any address, upon receipt of price.

C. M. SAXTON, Publisher,

25 Park Row, New York.

Books for Young Ladies.

I. WOMAN'S MISSION.
Gift Book for Young Ladies.

Or Woman's Mission; being Familiar Letters to a Young Lady on her Amusements, Employments, Studies, Acquaintances, male and female, Friendships, &c., by Dr. WM. A. ALCOTT, frontispiece on steel, Muslin, 307 pp., 12mo., 75 cents.

CONTENTS—LEADING HEADS.

General Views and Remarks; Spirit of Woman's Mission; Duties to Herself; Amusements; Employments; Studies, Books, &c.; Moral Character; Associates in the Family; Associates beyond the Family; Mere Acquaintance; Correspondents; Doing Good with the Pen; Particular Friendships: Society of the other Sex; Friendship with the other Sex; Qualifications for Friendship; Physical Qualifications; Seven Plain Rules; Disappointments; Doing Good; Pulling out of the Fire; Associated Effort; Church and Sabbath School; Truth, Justice, and Mercy; Labors among the Sick; Self-denial; Self-sacrifice.

*II. HEALTH, BEAUTY, AND HAPPINESS.
Young Woman's Book of Health.

By Dr. WM. A. ALCOTT, 312 pp., 12mo., Muslin, 75 cents.

The Young Woman's Book of Health, from the pen of Dr. Wm. A. Alcott, conveys, in simple and untechnical language, an amount of medical information which cannot but be of eminent service to those to whom it is addressed. The main object of the volume is to state the means of *preserving* one's health, rather than the way to hunt up one's health when lost. Doctor Alcott's reputation is well known in this country, and we feel satisfied that not only young females, but women of all ages, will find this his last book—eminently instructive and suggestive.—*Western Christian Adv.*

III. WHAT WOMAN MAY AND SHOULD BE.
Young Lady's Book.

Or, Principles of Female Education, by Rev. WM. HOSMER, frontispiece on steel, 301 pp. 12mo., Muslin, 75 cents.

CONTENTS.

Chapter I—Woman as a Human Being; Chapter II—Woman as a Social Being; Chapter III—Moral Education; Chapter IV—Intellectual Education; Chapter V—Physical Education; Chapter VI—Domestic Education; Chapter VII—Civil Education; Chapter VIII—Ornamental Education.

The foregoing works are eminently popular. They should occupy a place in the cabinet of every young lady in the land, as their counsels will always be found reliable, and their instructions possess a charm which renders them not only useful, but very agreeable and entertaining companions.

Sold by all Booksellers. Mailed, *post-paid*, to any address, upon receipt of price.

C. M. SAXTON, MILLER & Co., Publishers,
25 Park Row, New York.

www.ingramcontent.com/pod-product-compliance
Lightning Source LLC
Chambersburg PA
CBHW020100020526
44112CB00032B/593